Essione Ribeiro Souza
Valtemnir Ribeiro

Influence of defoliation on carbohydrate production in grapevines

Essione Ribeiro Souza
Valtemnir Ribeiro

Influence of defoliation on carbohydrate production in grapevines

Carbohydrate content in grapevine leaves in the Brazilian semi-arid region

ScienciaScripts

Imprint

Cover image: www.ingimage.com

This book is a translation from the original published under ISBN 978-3-330-99536-9.

Publisher:
Sciencia Scripts
is a trademark of
Dodo Books Indian Ocean Ltd. and OmniScriptum S.R.L publishing group

120 High Road, East Finchley, London, N2 9ED, United Kingdom
Str. Armeneasca 28/1, office 1, Chisinau MD-2012, Republic of Moldova, Europe
Managing Directors: Ieva Konstantinova, Victoria Ursu
info@omniscriptum.com

Printed at: see last page
ISBN: 978-620-8-58161-9

SUMMARY

CHAPTER 1

GROWING VINES

1. Vine characteristics

The Vitis genus is part of the Vitaceae family, which includes more than 90 species, of which those of American (Vitis labrusca L.) and European (Vitis vinifera L.) origin stand out for their economic and social importance (SOUZA & MARTINS, 2002).

The grapevine is a perennial, woody, deciduous and twining plant with a specialised support organ (tendrils) (QUEIROZ- VOLTAN & PIRES, 2003), with a twisted trunk, flexible branches, large, divided leaves, greenish flowers and fleshy fruit, called berries and clustered in bunches (GIOVANNINI, 1999). The bunches are formed by the peduncle and the stalk, at the ends of which the pedicels are formed, to which the fruit is attached. The bunches come in various shapes, the most common being conical, cylindrical or racemose (KUNH, 2003), which vary in size and are compacted or loose, depending on the number of berries attached and the characteristics of the cultivar (QUEIROZ- VOLTAN & PIRES, 2003).

The berries also have different shapes (ovoid, round, elongated, curved or flattened) and are made up of the pulp and the skin which, as well as enveloping the pulp, contains pigments that are responsible for the colour of the fruit. Inside the pulp are the seeds and if there are no seeds, the grapes are called apyrenic and are preferably marketed as table grapes (KUNH, 2003).

1.2. Economic aspects

Grape growing is of great economic and social importance in the north-eastern region of Brazil, in the Petrolma-PE/Juazeiro-BA cluster, where production spans 12 months of the year. In 2011 there was a 13 per cent increase in grape production in Brazil and the biggest increase in production was in the state of Pernambuco (24 per cent) as a result of the increase in the area planted in 2010. However, in 2011, it was in this state that there was the biggest reduction in the area planted (20 %), which led to a reduction in production in 2012. In Bahia, both grape production and the area planted with vines were reduced by 16 % and 15 % respectively (MELLO, 2012). Grapevine cultivation is responsible for most of the direct and indirect employment among irrigated crops in the region, generating a large annual turnover (AGRIANUAL, 2012).

The São Francisco Valley is located in Brazil's semi-arid tropics at 9°S latitude and 40°W longitude with an altitude of around 350 metres. It has average climatic indicators of 500 mm of annual rainfall, concentrated between December and March, an average temperature of 26°C and 50% relative humidity (IBRAVIN, 2012). The high evapotranspiration rates and water deficiency practically every month of the year favour growing vines under irrigation. It is Brazil's main tropical wine-growing region, with around 10,500 hectares of vineyards, distributed between the states of Pernambuco and Bahia (IBRAVIN, 2012), which account for 86 per cent of national production.

The region's production structure is made up of small producers, linked to colonisation projects and associated in cooperatives, and medium and large producers who operate on a corporate scale. Approximately 95 per cent of the area planted with vineyards is used to produce table grapes, mainly for the foreign market (IBRAVIN, 2011).

Viticulture in the middle of the São Francisco Valley is concentrated on the V. vinifera L. species, with the following cultivars standing out: Itália, Benitaka, Brasil, Red Globe, Festival, Thompson Seedless and Crimson Seedless (LEÃO, 2010). With yields of around 50 t ha^{-1} Italy and its mutations; 50 t $ha^{(-1)}$ Red Globe and 25 t h^{-1} the cultivars Festival, Thompson Seedless and Crimson Seedless (LEÃO et al., 2009).

The Submedio do São Francisco, although a recent wine-growing region, accounts for approximately 15% of fine wine production in Brazil (PEREIRA et al., 2009). The grapes are grown using trellis and espalier systems, with spacing between plants of between 1.00m and 2.00m and between rows of 3.00m for the trellis system, while for the espalier system, the spacing between rows has varied between 2.20m and 3.00m and between plants, 1.00m is used (LEÃO & SOARES, 2009), with the espalier system being better because it allows for mechanised cultivation and reduces labour costs (PEREIRA et al... 2009), 2009).

For an adequate alcoholic volume in the wine (at least 12%) the grapes must be harvested with more than 20°Brix (LIMA et al., 2009). This is more easily achieved in semi-arid Brazil, due to the intense sunshine, than in temperate zones, making it unnecessary to chaptalise (add sugar) during fermentation.

Lima et al. (2009) observed that in the São Francisco Valley, for grapes with seeds, full ripeness is reached between 120 and 130 days after pruning. According to Pereira (2009), the main varieties used to make red wines are Vitis vinifera L. Syrah, Cabernet Sauvignon, Tempranillo, Ruby Cabernet and Alicante Bouschet. For whites, the most commonly used varieties are Chenin Blanc, Moscato Cannelli, Sauvignon Blanc, Viognier and Verdejo. For sparkling wines and muscatel wines, the cultivars Italia and Moscato Cannelli are used, while the Syrah variety is used to make rosé wines.

The grape juice produced in the region has often been cited for its positive attributes, due to its high anthocyanin content, and is valued by the Brazilian consumer market. The domestic grape juice market has been growing at an average rate of 15 to 20% per year, with natural and/or whole grape juices growing at an even higher rate of around 40%. In two years, the sale of natural and whole grape juice increased by 86% in Brazil, from 13.7 million litres in 2007 to 25.5 million litres in 2009 (CAMARGO et al., 2008; IBRAVIN, 2013).

The main varieties used in the region for juice production are 'Isabel' on IAC 766 'Campinas' and IAC 572 'Jales' rootstocks, including the red grape cultivar BRS Cora, to obtain excellent quality juices with the right colour, flavour, aroma and sweetness (CAMARGO et al., 2008; IBRAVIN, 2013). The region therefore has favourable conditions for producing table grapes, wine and juice.

1.3. Environmental conditions for growing vines

Depending on the growing region, the grapevine has a succession of phases that characterise its growth and development. In some temperate climate regions, the vegetative cycle includes the phases of mobilisation and accumulation of reserves, vegetative growth and vegetative rest. In semi-arid tropical climates, predominant in the São Francisco Valley, it does not have dormancy or winter rest. The reproductive cycle of the branches developed during the vegetative cycle comprises the phases of flower induction, initiation and differentiation; flowering, pollination and fertilisation; fruit setting and development; fruit ripening and harvesting, characterised by the differentiation of ready buds and mixed buds formed in the previous cycle (Rodrigues et al. 2011).

The insertion of grapevines in different regions with tropical and subtropical climates has led to alterations in their physiological balance, promoting inexpressive and irregular shoots, causing low productivity. This is due to the adaptability of a species or cultivar to a production region, which is influenced by the interaction between its genetic makeup and the environment. Therefore, the behaviour of the same genotype can vary depending on the environment in which it is grown (PETRI et al., 2002).

Grapevines are characterised by their evergreen leaves, which maintain their photosynthetic capacity throughout the cycle. As a result of this characteristic, the leaves can withstand great seasonal variation in solar radiation and air temperature. These environmental factors will mainly influence the process of photosynthesis, which is the primary source of carbohydrates for all growth processes (MAGALHÃES FILHO, 2009).

When vines are grown under high temperature conditions, there is an increase in the concentration of sugars and a reduction in malic acids in the must, affecting the quality of the grapes (TEIXEIRA, 2009; KELLER, 2010), which, along with tartaric and citric acids, is the main organic acid responsible for the flavour of the grape, thus favouring the production of quality table grapes (WINKLER et al., 1974; COOMBE, 1987). Tartaric and malic acids are present in greater quantities within the vacuole of berry cells and are responsible for more than 90 per cent of total acidity (POSSNER & KLIEWER, 1985). Malic and tartaric acids stand out from citric acids and can vary depending on the cultivar and are formed in both green berries and leaves (GIOVANNINI, 2008).

A temperature between 26.8° and 32.2°C is the optimum range for development and has a strong influence on the formation and development of the pollen tube, as it influences the rate of cell division and elongation (MOURA et al., 2009). Flowers begin to develop when the average daily temperature reaches 18°C. In tropical climates, the dormant period is achieved through water management during the rest period, favouring production at any time of the year (ARAÚJO, 1994).

During the resting phase, the irrigation rate is reduced taking into account the use of crop coefficients (Kc), which relate the water consumption of the reference crop (grass) to the specific crop at a given stage of its development. At this stage, Kc 0.4 is used so that the plant can continue its photosynthetic activity and thus produce and accumulate carbohydrates. However, 15 days before the application of ethylene, this rate is reduced and most of the time set to zero, in order to maximise the effect of ethylene and thus promote leaf fall. And in the pruning phase, it is irrigated again, taking into account Kc 0.7, to stimulate sprouting.

Air temperature also interferes with the photosynthetic activity of plants, since it involves biochemical reactions whose catalysts, the enzymes, are dependent on temperature to express their maximum activity. Photosynthesis reactions are less intense at temperatures below 20°C, intensifying as the temperature rises and reaching a maximum between 25 ° and 30°C, reducing again when the temperature exceeds these values and ceasing when it approaches 45°C. The average temperature range considered ideal for grapevines is between 20° and 30°C (COSTACURTA & ROSELLI, 1980), although in some regions the average temperature range considered ideal is between 25° and 30°C (FERREIRA et al., 2004; LIMA FILHO et al., 2009; SOUZA et al., 2010).

Solar radiation is the main source of energy for the processes of photosynthesis and evapotranspiration in the grapevine crop, as well as leaf area, the type of trellising system and the cultural management employed (TEIXEIRA et al., 2007). Vapour pressure differences between the plant surface and the air are another important factor in evapotranspiration rates. Wind speed also plays a role in the evapotranspiration process, because this process depends on air turbulence, which is affected by plant architecture (TEIXEIRA et al., 2008).

The water vapour pressure deficit between the vine canopy and the surrounding air is a determining factor in vapour removal. Well-irrigated vines in arid regions consume large quantities of water due to the intense availability of solar energy and the desiccating power of the atmosphere (ALLEN et al., 1998).

In some regions, grapevine water requirements vary from 500 to 1,200 mm, depending on the climate, the length of the phenological phase (SENTELHAS, 1998), the rootstock and the soil. Both water deficiency and excess significantly affect the behaviour of the phenological stages of the grapevine crop, compromising fruit quality and productivity (TEIXEIRA et al., 2002). Water deficiency, when it occurs during the initial period of berry growth, causes a reduction in fruit size; when it occurs during ripening, it delays ripening, affecting colour and favouring fruit burning by solar radiation. In the final stage of ripening, the vine's water consumption decreases.

Excess water, combined with high temperatures, makes the crop very susceptible to diseases, especially powdery mildew and downy mildew. For good productivity, it is recommended that water needs are met in accordance with the crop's water requirements (TEIXEIRA & AZEVEDO, 1996).

1.4. Vine phenology

There are various physiological and environmental factors that influence the development and growth of grapevines grown in irrigated areas in the north-east of Brazil, defining the development of phenological stages that differ from other growing regions, which is why the pruning season marks the start of the new phenological cycle, which will be conditioned by the prevailing climatic factors during the period. It is therefore necessary to understand and know the phenology of the grapevine.

When introducing new varieties, phenology plays an important role, as it allows the duration of the vine's development phases to be characterised in relation to the climate, as well as being used to interpret how the different climatic conditions interact with the crop (TERRA et al., 1998). This allows management practices to be developed in the different producing regions, especially in semi-arid regions, given that in this region the behaviour of the vine is different from traditional growing regions, as the productive and vegetative cycles are conditioned to irrigation control and pruning times (ASSIS & LIMA FILHO, 2000).

According to Sentelhas (1998), grapevines need between 1,200 and 1,400 hours of solar radiation over the course of a cycle, and this is greatest in the reproductive sub-period, depending on the region where the grape is grown and the cultivar.

The reproductive phase, flowering, is an important process in the grapevine cycle, as this is when fruit formation begins and it is highly influenced by environmental factors and cultivation practices (pruning, thinning and cluster pinching) (KELLER et al., 2001; VASCONCELOS et al., 2009). Environmental factors influence flowering by modifying the plant's chemical composition, particularly the endogenous hormonal balance, and also through their impact on photosynthesis (VASCONCELOS et al., 2009).

According to Dokoozlian (2002) increases in berry mass, volume or diameter during development are typically characterised by a double sigmoid curve, resulting from two consecutive phases of growth, separated by a phase of slow or zero growth and controlled by three factors: cell number, cell volume and content of organic solutes (sugars). The number of cells is established during the first three weeks after anthesis. The number of cell divisions prior to anthesis is the main determinant of the number of cells per berry. Cell volume increases significantly during phase I (phase of rapid berry growth), shortly after flowering; in this phase, the berries have a firm texture, green colour (presence of chlorophyll), low sugar content and accumulation of organic acids, remaining relatively constant during phase II (reduction in the rate of berry growth, growth of the embryo, increase in the concentration of organic acids to its maximum peak). After this phase, the berries remain firm, but begin to lose their chlorophyll and resume their rapid expansion (phase III - second phase of rapid berry growth and fruit ripening), the accumulation of sugars begins and the concentration of organic acids decreases.

1.5. Hormones and Plant Regulators

Plant hormones are organic, non-nutritive compounds that occur naturally in plants and act in small concentrations to alter different physiological responses, influencing germination, vegetative growth, flowering, fruiting, senescence and abscission (VIEIRA et al., 2010). According to Taiz & Zeiger (2013) plant hormones are chemical messengers produced in one cell that modulate cellular processes in another cell by interacting with specific proteins that function as receptors linked to signal transduction pathways.

Plant regulators are synthetic substances that alter the morphology and physiology of the plant and can lead to qualitative and quantitative changes in plant development. These substances can inhibit or promote physiological, morphological and histological changes, and can cause changes in the growth orientation of the aerial part, lignification and canopy density (BIEHNETAL, 1966; WINKLER, 1997 apud SPINELLI et al., 2010).

The different plant hormones act at different stages of plant development (MULLINS et al., 1992). Auxins regulate different stages of plant development, such as cell expansion and division, as well as the differentiation of vascular tissues (MOORE, 1989). The term auxin came from the Greek *auxein,* meaning "to increase" or "to grow", due to the fact that this substance promotes the growth of coleoptile sections of seedlings. The main auxin in plants is indole-3-acetic acid (IAA). Synthetic auxins such as indolylbutyric acid (IBA), naphthaleneacetic acid (NAA) and 2,4-dichlorophenoxyacetic acid (2,4-D) stand out (DINIZ, 2005; TAIZ & ZEIGER, 2013).

According to Cruz (2013) auxins are produced at the end of young or adult stems, young leaves, flowers, fruits and seeds. One of the main effects of auxins is to promote the elongation of newly formed cells from meristems, promoting the growth of stems and roots. The effect of auxins depends on their concentration in the plant, as the concentration that promotes growth in the stem can have the opposite effect. Therefore, very high concentrations can inhibit cell growth, as the sensitivity of cells to auxins varies according to the different organs of the plant.

Deluc et al. (2007), studying the concentration of auxins in the different stages of fruit development, observed that auxin concentrations were high in the initial phase of fruit development, decreasing after veraison, indicating the role of this phytohormone in promoting cell division and expansion during this stage. He also found that auxin concentrations remain relatively constant throughout fruit development. This can be explained by the fact that this phytohormone is produced in the pollen and after fertilisation, the growth of the fruit depends on the auxin produced in the growing seeds.

It is important to emphasise that acidification of the cell wall is not the only way in which auxin induces plant cell elongation. Auxin must affect other important processes related to cell growth, such as the absorption and production of osmotic solutes, as well as controlling the growth and maintenance of the cell wall structure. The absorption of solutes depends largely on the activity of the H^+ -ATPase, which is induced by auxin. Auxin also increases the activity of certain enzymes involved in polysaccharide biosynthesis. These polysaccharides can be used to synthesise new cell wall materials, contributing to continued cell growth. It is believed that plant responses to auxin involve both changes in protein activity (enzymes, ion channels, etc.) and gene expression (TAIZ & ZEIGER, 2013).

According to Deng et al. (2007) the abscission of grape berries is associated with an increase in the activities of hydrolases in the abscission zone, particularly the enzymes cellulases, pectinases and polygalacturonase. According to Taiz & Zeiger (2013) ethylene is the main plant hormone regulating the process of abscission, with auxin acting as a suppressor of the effect of ethylene.

According to Hespanhol-Viana et al. (2007) the adherence of the berries to the pedicel also depends on the ripeness of the grape and the presence or absence of seeds. Most apirenic grapes are characterised by the absence or reduced presence of the "brush", which is why they are more genetically predisposed to the separation of the berry from the pedicel. Colapietra et al. (1995) found a notable difference between seeded and apyrenic grapes in terms of the load required to separate the berry from the pedicel.

Gibberellins are the most widely used group of plant regulators in commercial viticulture, mainly due to their effects on increasing the size of grape berries, resulting from their action on increasing cell division and expansion (FILIIO et al., 2009; RODRIGUES et al., 2011). The effect of **GA3** varies depending on the cultivar, rootstock, concentration, method and time of application and environmental conditions (TECCHIO et al., 2009).

Elongation, expansion and cell division depend on the action of auxins and gibberellins, both of which act in different processes, but together contribute to the whole process. Thus, auxins and gibberellins stimulate growth through both expansion and cell division (ROSS & O'NEILL, 2001).

Cytokinins act in various phases of plant development, such as cell division, stem initiation and growth, delaying leaf senescence, photomorphogenesis, breaking apical dominance, fruit development, hydrolysis of starch reserves, increased stomata opening, chloroplast development, nutrient mobilisation and induction of parthenocarpy in fruit (COLL et al., 2001; DAVIES, 2004). Their main biosynthesis site is the roots, from where they are transported via the xylem to the stem.

Cytokinins are important in controlling many aspects of the vine, including flower differentiation, bunch growth, berry set and development, ovule growth and the transformation of tendrils into fertile inflorescences (SRINIVASAN & MULLINS, 1980; KANELLIS & ROUBELAKIS- ANGELAKIS, 1993; RUIZ, 1998).

The group of synthetic cytokinins is divided into two, purine synthetic cytokinins, by modifying the side chain at the N-6 position of the adenine base, one of the most active being PBA [(6- benzylamino)-9-(2-tetrahydropyramyl)-9H-purine] and the non-purine synthetic cytokinins derived from urea such as diphenylurea, CCPU (N-(2-chloro-4-pyridyl-N-phenylurea)) or forchlorfemuron, benzimidazole, imidazole, fluorophenylbiuret, TDZ (N-phenyl-N-1,2,3-thidiazol-5-thiourea). This large number of synthetic cytokinins are widely used in most plant species because they have significant effects on fruit production and quality.

Pires et al. (2003) found in bunches of 'Centennial Seedless' that CPPU applied at doses of 0; 5.0; 7.5; 10.0; 12.5; 15.0 and 17.5 mg L-1 at 14 days after full bloom, increased the mass and width of the bunches; the mass, length and width of the berries; the diameter of the pedicels and reduced the soluble solids content. They also observed that the highest doses of CPPU resulted in very large bunches, excessively compacted, with rigid and very thick pedicels. These disturbances caused to the bunch by the application of the highest doses of CPPU, as well as increasing the amount of labour required to decompress the bunches, favoured the appearance of hard berries, hindering the ripening process and disqualifying them from being marketed.

Cytokinins influence the mobilisation of nutrients from other parts of the plant, which are translocated in the phloem and move from a place of production or storage (source) to a place of use or storage (drain). Thus, the nutritional condition of the plant regulates cytokinin levels and, in turn, the relative ratio between high levels of auxin and low levels of cytokinins positively affects the rate of root growth (TAIZ & ZEIGER, 2013). Various sources of plant regulators have been used in different crops, applied alone and/or in combination, with the aim of improving fruit size and quality, colour, senescence, chemical thinning, uniform sprouting, reducing degranulation, post-harvest conservation, fruit storage, seed germination, overcoming dormancy, floral induction, adjustment between male and female flowers, etc, including gibberellic (GA), naphthaleneacetic (NAA), abscisic (ABA) and indolylbutyric (IBA) acids, 6-benzyladenine (BA) and ethephon, etc.

1.6. Foliar fertilisation

Nutrients play important roles in plants: nitrogen (N) is the most abundant mineral element in plants, as it is a constituent of amino acids, proteins, enzymes, coenzymes, chlorophylls, nucleic acids, nucleotides and other important compounds in cellular metabolism (MARENCO & LOPES, 2011), assisting in the formation of leaves, favouring rapid plant growth, the formation of chloroplasts, as well as transporting amino acids (glutamate and aspartate) and amides (glutamine and asparagine).

Asparagine and glutamine link carbon and nitrogen metabolism. It acts not only as a protein precursor, but as a key element in the transport and storage of nitrogen, due to its stability and high nitrogen: carbon ratio. The main route for asparagine synthesis involves the transfer of the amide nitrogen radical from glutamine to asparagine, and this reaction is catalysed by asparagine synthetase (AS). High levels of light and carbohydrates, conditions that stimulate glutamine synthetase (GS) and glutamate synthetase (Fd-GOGAT) in the plastids, inhibit the expression of the genes that code for AS and their activity. The antagonistic regulation of these competitive pathways helps to balance carbon and nitrogen metabolism in plants. Conditions of ample energy, i.e. high levels of light and carbohydrates, stimulate GS and GOGAT, inhibiting AS and thus favouring the assimilation of nitrogen into glutamine and glutamate, rich compounds that participate in the synthesis of new compounds in plants (TAIZ & ZEIGER, 2013).

The mechanism according to which the level of amino acids in the root phloem regulates the absorption and assimilation of nitrogen by the plant was suggested by Imsande & Touraine (1994). This mechanism is based on the observation that, during rapid vegetative growth, the rates of nitrate reduction and amino acid synthesis in the leaves are high. There, most of the amino acids are used to synthesise chlorophyll, the Rubisco enzyme and other proteins and, as a result, the level of amino acids in the phloem that are translocated to the roots is low. On the other hand, during the reproductive phase there is a reduction in the nitrate rate; at the same time, due to the remobilisation of leaf N for the development of the inflorescences, the export of amino acids from the leaves increases, enriching the phloem with these compounds, which will be translocated to the roots. The proposed mechanism suggests that these amino acids cause a reduction in the rate of **$NO3^-$** absorption (IMSANDE & TOURAINE, 1994).

Phosphorus (P) is part of the phospholipid molecule (an essential constituent of cell membranes), phosphate sugars, nucleotides (NAD, FAD, ATP, UTP, GTP), coenzymes (CoA, FAD) and nucleic acids (DNA, RNA). Of the total inorganic phosphate (Pi) in plants, 85 to 90 per cent is located in the vacuole, also called the non-metabolic compartment (MARSCHNER, 2011). As well as taking part in respiration, energy transfer, cell division and cell growth, it stimulates root growth and formation (MARENCO & LOPES, 2011).

The majority of sulphur (S) is translocated via the xylem to the aerial part, where it is reduced and assimilated (ZHAO et al., 1999; LEUSTEK et al., 2000), with a minority being metabolised in the root.

Most of the S in the plant is found in proteins, as part of the amino acids, cystine, cysteine (where S enters organic compounds) and methionine (where ethylene synthesis begins). It is also present in vitamins, thiamine (vitamin **B1**) and biotin (vitamin H); in coenzyme A (CoA); it participates in respiration and in the synthesis of fatty acids, sterols and plant hormones (gibberellin, abscisic acid, ethylene, brassinosteroids and cytokinins), although none of these plant hormones contain this element. It is also a constituent of S-adenosylmethionine, which donates its methyl group during the formation of steroids and lignins. It is also essential for ferredoxin (sulphur reducing agent in green leaves), an important Fe-S protein in photosynthesis; and glutathione, a tripeptide that acts as an antioxidant in various detoxification processes (DROUX, 2004). When bound to Ca^{2+}, it favours its migration to the subsoil, attracting roots. Sulphur is also involved in chlorophyll synthesis (MALAVOLTA, 2006).

Molybdenum is the least abundant micronutrient in the soil and plant, but it is extremely important for plant metabolism (MALAVOLTA, 1980). Molybdenum ions (Mo^{+4} to $Mo^{(+6)}$)) are cofactors of various enzymes, including nitrate reductase and nitrogenase (TAIZ & ZEIGER, 2013). It is also closely related to electron transport during biochemical reactions in plants (PRICE et al., 1972; LANTMANN, 2002).

Cobalt (Co) is a chemical element necessary for the synthesis of cobalamin (vitamin **B12**), which participates in the metabolic steps for the formation of leghemoglobin, which has a high affinity for oxygen and regulates the concentration of this gas in legume nodules, preventing the inactivation of the nitrogenase enzyme (FAVARIN & MARINI, 2000).

The plant's nutritional condition also regulates cytokinin levels and, in turn, the relative ratio between cytokinin and auxin determines the relative growth rate of roots and aerial parts, i.e. high concentrations of cytokinin promote aerial part growth and, conversely, high levels of auxin promote root growth. In the presence of low nutrient levels, cytokinin levels are also reduced, resulting in increased root growth, allowing the plant to more efficiently acquire the nutrients present in the soil. In comparison, soils with optimal nutritional conditions promote an increase in cytokinin levels, favouring the growth of the aerial part and thus maximising photosynthetic capacity (TAIZ & ZEIGER, 2013).

1.7. Gas exchange

Photosynthesis and, consequently, respiration depend on a constant flow of **CO_2** and **O_2** in and out of the cell; this free flow is a function of the concentration of **CO_2** and **O_2** in the intercellular spaces, which depend on stomatal opening, the main controller of gas flow (MESSINGER et al., 2006).

Stomatal conductance, in turn, can be understood as the physiological mechanism that vascular terrestrial plants have for controlling transpiration (MESSINGER et al., 2006). According to Naves-Barbiero et al. (2000), the epidermis of leaves is surrounded by a cuticle that is relatively impermeable to both water vapour and carbon dioxide and contains a variable number of stomata, whose response, through the regulation of stomatal conductance, controls leaf transpiration and is influenced by light, the concentration of atmospheric carbon dioxide, humidity and temperature.

The reduction in photosynthesis may be due to limitations in the supply and utilisation of **CO_2**, light energy and phosphate (FARQUHAR & SHARKEY, 1982). Under natural conditions, several environmental variables that are potentially related to photosynthesis change concomitantly and frequently, and it is difficult to identify the individual effect of each one in field experiments (MAGALHÃES FILHO, 2009).

Photosynthesis is essential for plant survival and is the main carbon supply process for plant growth and development (DORNHOFF & SHIBLES, 1970). It requires more than half of the nitrogen (N) in the leaf from the rubisco enzyme and is therefore greatly affected by N availability (DIETZ & HARRIS, 1997; LAMBERS et al., 1998). Leaf N concentration correlates positively with leaf chlorophyll content, which in turn correlates with photosynthesis and the latter with crop productivity (SMEAL & ZHANG, 1994; SCHADCHINA & DMITRIEVA, 1995).

1.8. Chlorophyll pigments

Chlorophyll pigments are present in higher plants as chlorophylls a, b, c, d and e, with chlorophylls a and b being the most abundant and therefore the most important. These are constantly synthesised and destroyed, and their processes are influenced by factors both internal and external to the plants. Among the external factors, mineral nutrients stand out, as they integrate the molecular structure of these pigments, as well as acting in some stage of the reactions that lead to the synthesis of these pigments (TAIZ & ZEIGER, 2013).

The pigments involved in photosynthesis are chlorophylls a and b, carotenoids and phycobilins (KLUGE, 2004). There are several types of chlorophyll, a, b, c1, c2 and d, but chlorophyll a is the most important and the others are called accessory pigments. Chlorophyll a is the pigment responsible for the photochemical process of photosynthesis, while the other pigments help to absorb light and transfer radiant energy to the reaction centres (TAIZ & ZEIGER, 2013). One of the differences between chlorophyll b and chlorophyll a is due to the linkage of carbon 3 of the pyrrole ring, with chlorophyll a having the methyl radical while b has the aldehyde radical (VON ELBE, 2000; LARKUM & KÚHL, 2005).

Chlorophylls are responsible for the green colour of plants and have low light absorption in the region of the electromagnetic spectrum corresponding to this colour. Chlorophylls are essential for plant development and are synthesised in a characteristic biosynthetic pathway that involves the need for more than one gene (THIMM et al., 2004).

The quantity and proportion of chlorophylls in the leaves are closely related to the plant's nutritional status. Chlorophyll content can therefore be used as an index to diagnose nutritional status (KAWASHIMA & NAKATANI, 1998). Chlorophyll is constantly synthesised and destroyed (photo-oxidation) in the presence of light, but at very high light intensities the speed of decomposition is greater and equilibrium is established at a lower concentration (KRAMER & KOZLOWSKI, 1979).

According to some authors, shaded plants have a higher relative proportion of chlorophyll b (ENGEL, 1989; ENGEL & POGGIANI, 1990; 1991), as it enables energy from other wavelengths to be captured and transferred to a specific molecule of chlorophyll a, which effectively takes part in the photochemical reactions of photosynthesis (WHATLEY & WHATLEY, 1982), which may be related to an increase in the proportion of the chlorophyll a/b collector complex (THORNBER, 1975). In some species, the degradation of chlorophyll b is more difficult than that of chlorophyll a (ENGEL & POGGIANI, 1991).

Chlorophylls are closely related to the photosynthetic efficiency of plants and, consequently, to their growth and adaptability to different environments (ALMEIDA et al., 2004). Chlorophyll and carotenoid levels in leaves are used to estimate the photosynthetic potential of plants due to their direct link to the absorption and transfer of light energy and to growth and adaptation to different environments. A plant with a high chlorophyll content is capable of achieving higher photosynthetic rates due to its potential value for capturing quanta of light in a unit of time (PORRA et al., 1989; CHAPPELLE & KIM, 1992).

1.9. Lipid peroxidation and antioxidant enzymes

During normal metabolic activity, such as electron transport in chloroplasts, mitochondria and the plasma membrane and/or as a result of environmental disturbances (extreme temperatures, drought, salinity, ultraviolet radiation, ozone, diseases, herbicides, pollutants, etc.), reactive oxygen species (ROS) can be formed. When these ROS are produced in excess, overcoming the action of the antioxidant system, oxidative stress occurs (KIM et al., 2004; SCANDALIOS, 2005). There are various biochemical indicators of oxidative stress, including the rate of lipid peroxidation.

The production of EROS can initiate the process of lipid peroxidation in cell membranes, forming lipid hydroperoxides. Peroxidation of membrane lipids is one of the most significant events in oxidative stress, as it leads to a reduction in fluidity, changes in ion permeability and other functions associated with membranes (QUEIROZ et al., 1998). The biochemical events resulting from this reduction in membrane fluidity include interference in the functions of proteins, a reduction in energy supply, loss of compartmentalisation, accentuated release of ions and other events that disrupt normal metabolism and lead to imbalance and loss of essential functions (AZIZ & LARHER, 1998).

To minimise the deleterious effects of ROS and modulate their quantity, aerobic organisms have developed antioxidant defence systems, also known as scavengers, enzymatic and non-enzymatic. Non-enzymatic defences include vitamins C and E, glutathione (GSH), 0-carotene, phenolic compounds, tocopherols and polyamines. The enzymatic defence system involves the enzymes superoxide dismutase (SOD), catalases (CAT), peroxidases (POD), glutathione peroxidase (GPX), ascorbate peroxidase (APX), glutathione reductase (GR) and glutathione S-transferase (GSTs) (BLOKHINA et al., 2003; SCANDALIOS, 2005). Alongside other physiological mechanisms, the efficiency of the antioxidant system increases the plant's capacity for tolerance, due to the reduction in the effects caused by ROS (GIANNAKOULA et al., 2010).

Superoxide dismutase (SOD) is a metalloprotein that catalyses the dismutation of **O2**$^{\circ-}$ into **O2** and **H2O2**. Three classes of SOD are known in plants and can be linked to a metal (Cu/Zn, Mn and Fe). Plants normally have Cu/Zn-SOD in the cytosol, Cu/Zn and/or Fe-SOD in the chloroplast and Mn-SOD in the mitochondria. The three enzymes have distinct molecular properties, including differential sensitivity to the inhibitor, and are localised in different subcellular compartments. SODs are considered important antioxidant agents, but at high concentrations in animal and bacterial cells they can induce dysfunction and cell death (BAKER & ORLANDI, 1995), as well as being a powerful catalyst for increasing reaction speed.

Catalase (CAT) is the most active enzyme produced by nature and converts H2O2 into **H2O** and **O2**. Plants have several isoforms of catalase, which are present in peroxisomes and glyoxysomes. They are the main H2O2 detoxification enzymes in plants and can directly dismutate H2O2 or oxidise substrates such as methanol, ethanol, formaldehyde and formic acid. Catalases can be divided into three classes: the first removes H2O2 produced during photorespiration in photosynthetic tissues; the second is produced in vascular tissues and may have a lignification function, but its exact biological function remains unknown; and the third is abundantly present in seeds and young plants, whose activity is related to the removal of H2O2 produced during the degradation of fatty acids in the glyoxysome. Catalases function as a cellular H2O2 scavenging channel (BREUSEGEM et al., 2001).

The group of peroxidases (POD) includes enzymes (EC 1.11.1.7) capable of catalysing the transfer of hydrogen from a donor to **H2O2**. In plants, the action of this group of enzymes constitutes antioxidative protection. Peroxidase activity can increase in plants subjected to various types of stress (SIEGEL, 1993). Peroxidases participate in various physiological processes, catalysing the oxidation and eventual polymerisation of hydroxycinnamic alcohol in the presence of hydrogen peroxide, giving rise to lignin, an important physical plant defence mechanism. This enzyme is involved in various reactions in the plant, polysaccharide bonds, indole-3-acetic acid (IAA) oxidation, monomer bonds, lignification, wound healing, phenol oxidation, pathogen defence, regulation of cell elongation and others (GASPAR et al., 1982; KAO, 2003). According to Pascholati & Leite (1994), peroxidase is directly linked to tissue lignification, polymerising lignin by oxidising the hydroxyls of phenolic groups.

The degree of peroxidation of membrane lipids generates malonic aldehyde (MDA), which is a product of the decomposition of fatty acids in biomembranes and is a way of attesting to the presence of free radicals in metabolic routes that involve electron transfer reactions, such as in the photochemical phase of photosynthesis (MITTLER, 2002). These radicals, when produced in excess, can be destructive to the cell by reacting with the unsaturated fatty acids of membrane phospholipids and altering their functionality, promoting lipid peroxidation (SCANDALIOS, 2005).

According to CAKMAK & HORST (1991), the reduction in the activity of some enzymes, such as Catalase, indicates that, in some plants kept under stress conditions, the H2O2 produced may be consumed more in oxidative processes, such as lipid peroxidation, than eliminated from the metabolism by the action of antioxidant enzymes.

Several studies in the literature have reported lipid peroxidation and the activities of the antioxidant enzymes SOD, CAT and POD in plant tissue in response to different stresses, such as Queiroz et al. (1998) studying coffee crops; Fang et al. (2001) and Queiroz et al. (2002) working with aroeira-do-sertão; Kang & Saltveit (2002) working with cucumber seedlings; Da Costa & Huang (2007) with the species Agrostis capillaris L. and A. stolonifera L. and Souza et al. (2011) studying rice cultivation.

1.10. Carbohydrates

Sugars from photosynthesis act as a substrate for energy metabolism and carbohydrate biosynthesis, providing conditions for the growth and development of plant tissues. In addition, sugars can act as secondary messengers, ensuring that the plant continues to develop even after biotic or abiotic stresses (HAMMOND & WHITE, 2008), as well as directly regulating gene expression through signalling networks. Sugar in abundance promotes growth and the storage of carbohydrates in storage drains and when the rate of photosynthesis is high, total soluble sugars and starch accumulate in the leaves (TAIZ & ZEIGER, 2013). According to Palliotti & Cartechini (2001), the amount of carbohydrates in grapevines is directly related to their phenological stage and seasonality. Majerowicz (2004) points out that in the metabolism of carbon in leaves, starch acts as a regulator between demand (mobilisation to growing and reserve tissues) and source (photosynthesis).

Santos et al. (2013), studying the metabolism of the 'Syrah' grapevine under three irrigation strategies (controlled deficit irrigation, deficit irrigation and full irrigation) concluded that total soluble sugars and reducing sugars increased in all irrigation strategies at the end of the cycle and that it is possible to reduce water consumption in wine crops in the semi-arid northeast without significantly affecting the metabolism of proteins and sugars.

During the ripening phase, sucrose is converted into monosaccharides that gradually accumulate in the berries (ALI et al., 2010). The levels of glucose, fructose and sucrose, as well as the activity of carbohydrate metabolising enzymes in grapevine leaves, are influenced by seasonal and phenological variations in the vines (PALLIOTTI & CARTECHINI, 2001).

According to Santos et al. (2010), the synthesis and accumulation of organic and/or inorganic solutes varies between plant species, as well as between different cultivars of the same species.

CHAPTER 2

INFLUENCE OF DEFOLIATION ON CARBOHYDRATE PRODUCTION IN GRAPEVINES

Viticulture is practised in various regions of the world, with Brazil occupying 15th place.

The Submedio São Francisco Valley (Juazeiro, BA and Petrolina, PE) is home to the largest table grape production areas in the country and stands out on the national scene, not only because of the expansion of the cultivated area and the volume of production, but mainly because of the quality of the grapes produced (Correia & Silva, 2004).

In 2010, Brazil produced approximately 57 per cent of table grapes and 43 per cent of fine grapes for wine and grape juice production (Mello, 2011).

The São Francisco Valley sub-medium region is the most representative territorial area, with 10,000 ha under vine, accounting for 95 to 98 per cent of Brazil's fruit exports (AGRIANUAL, 2012).

The grapevine is grown under different environmental conditions in various regions of the world. In semi-arid tropical climates, it behaves in a completely different way to subtropical and temperate climates, and is dependent on irrigation control and pruning times. It can be said that climatic conditions influence the phenology and physiology of plants and the production and quality of grapevine fruit (Albuquerque & Albuquerque, 1982).

The grapevine (Vitis vinifera L.) is a species that is often subjected to extreme environmental situations in the climates in which it grows, such as the semi-arid tropical climate. Water deficit, combined with high temperatures and radiation during the summer season, and a high vapour pressure deficit reduce gas exchange in the leaves and negatively affect the quantity and quality of production. In grapevines, as in many other species, the closure of stomata is the plant's first response to water deficit under field conditions (Chaves et al. 2002).

The decrease in photosynthesis that occurs in grapevines under conditions of water deficit can also reduce the production and accumulation of carbohydrates and, consequently, affect their growth and production (Souza, 2004).

Photosynthesis is the storage of solar energy by plants. The photons absorbed excite chlorophyll molecules, which, when excited, release this energy as heat, fluorescence, energy transfer or photochemistry (Taiz & Zeiger, 2009).

According to Pommer et al (2003), photosynthetic activity is closely correlated to intrinsic plant factors, such as vigour and, as a result of the number of leaves, leaf area, and extrinsic factors, those that influence the part of the leaves that receive sunlight, such as the conduction system and the microclimate around the plant. Until they reach a third of their final size, grapevine leaves consume more substances than they produce, and the photosynthesis balance will turn positive when the leaves stop importing sucrose. Synthesis reaches its maximum value around forty days after the leaves emerge, i.e. with the greatest development of the leaf limb

(Pommer et al, 2003).

Photosynthesis and, consequently, respiration, depend on a constant flow of **CO2** and **O2** in and out of the cell; this free flow is a function of the concentration of **CO2** and **O2** in the intercellular spaces, which depend on stomatal opening, the main controller of gas flow (Messinger et al., 2006).

Stomatal conductance, in turn, can be understood as the physiological mechanism that vascular terrestrial plants have for controlling transpiration (Messinger et al., 2006). According to Naves-Barbiero et al. (2000), the leaf epidermis is surrounded by a cuticle that is relatively impermeable to both water vapour and carbon dioxide and contains a variable number of stomata, whose response, through the regulation of stomatal conductance, controls leaf transpiration and is influenced by light, atmospheric carbon dioxide concentration, humidity and temperature. According to these authors, stomatal conductance is proportional to transpiration, net photosynthesis and leaf water potential.

Based on their knowledge of the biochemical characteristics of Rubisco in requiring NADPH2 and ATP for **CO2** assimilation, Farquhar et al. (1980) developed a model for photosynthesis in C3 plants. The model demonstrates the basic principles of the biochemistry of photosynthesis, explaining the physiological properties of photosynthesis in intact leaves. Further refined by Von Caemmerer (2000) and Sharkey et al., (2007), the photosynthetic rate can be limited by three factors: 1° by the amount and/or activation state of Rubisco; 2° by the regeneration of ribulose-1,5-biphosphate (primary CO_2 acceptor); 3° by the chloroplast's ability to exchange triose phosphate for inorganic phosphate (Pi) with the cytoplasm (Von Caemmerer, 2000). The biochemical model of Farquhar et al. (1980) is widely used in plant ecophysiology research and more recently in modelling related to global climate change. Fundamentally, however, the model allows an assessment of how factors that could potentially affect the photosynthetic process, such as biotic and abiotic stress factors, can be distinguished between stomatal and non-stomatal effects. Stomatal effects are directly related to the restriction on gas exchange imposed by stomata, while non-stomatal effects can be divided into the carboxylative activity of Rubisco, the electron transport flow associated with the RuBP regeneration process and the utilisation of phosphate trioses.

In turn, photorespiration is one of the most important physiological characteristics that differentiates C3 plants from C4 plants. It involves the loss of carbon by the leaves, simultaneously with the process of **CO2** absorption by photosynthesis. It is a process of releasing **CO2** into the light, functionally and metabolically linked to photosynthesis. In fact, photosynthesising tissues release CO2 more intensely in the light than in the dark. Illuminated leaves, in conditions of equilibrium in terms of **CO2** exchange with the environment, release a high amount of **CO2** in the first few minutes following the start of the dark period. This post-illumination increase in **CO2** release seems to mean that some compound synthesised in the light is stored, and its use continues to occur in the dark, as long as there is substrate available for the process (Taiz & Zeiger, 2009).

Chlorophylls are the most abundant natural pigments present in plants. They are responsible for converting light radiation into energy and occur in the chloroplasts of leaves and other plant tissues. Studies on a wide variety of plants have shown that the chlorophyll pigments are the same. The apparent differences in plant colour are due to the presence and variable distribution of other associated pigments, such as carotenoids, which always accompany chlorophylls (Von Elbe, 2000).

In general, the increase in chlorophyll concentration as light intensity decreases increases the capacity to absorb light of different wavelengths from the photosynthesis peaks, such as light in the green range, which is present in large quantities in the interior of forests. In general, chlorophyll and carotenoids tend to increase in concentration as light intensity decreases (Ferraz & Silva, 2001; Fontes & Silva, 2001).

Various methods can be used to assess the nutritional status of plants, but their speed and safety must be taken into account. The SPAD-502, even with a compensation system for changes in leaf thickness and water content (light absorbed at 940 nm), shows large variations in readings between species (Richardson et al., 2002) and between genotypes (Jifon et al., 2005) of fruit plants grown in the same environment, due to differences in leaf structure and anatomy. Leaves of the same grapevine genotype can show large variations in the SPAD index, depending on the phenological stage, position/age of the leaf on the plant and year of assessment (Porro et al., 2001).

In particular, the grapevine plant is extremely demanding in terms of carbohydrate levels for the efficiency and quality of its fruit formation. Already in the first two to three weeks after budbreak, the movement of assimilates is essentially towards the tips of the branches, which at this stage depend on carbohydrates and nitrogen compounds stored in other parts of the plant (Pommer et al., 2003).

Photosynthetically active tissues, such as mature leaves, produce more carbohydrates than they need to maintain their metabolism and growth, exporting the surplus photoassimilates, in the form of sucrose, to less photosynthetically active or inactive tissues, such as young leaves, roots, bunches or branches. However, source-drain relationships are not static. During vegetative growth, most carbohydrates are transported to the roots and young leaves, while after flowering, carbohydrates are prioritised for the fruit and roots (Roitsch et al., 2003).

Among the various varieties of seeded grapes grown in the Vale do Submédio São Francisco region, the Itália or Piróvano 65 cultivar is the most representative of fine grapes in Brazil, and its production is concentrated in the states of São Paulo, Paraná, Minas Gerais and the Vale do São Francisco region in north-eastern Brazil (Pommer et al, 2003).

The domestic market favours muscat cultivars and, in the Vale do Submédio São Francisco region, the Itália cultivar and its Muscat clone are the main representatives. Because of its desirable characteristics, despite the difficulties involved in growing it, it has expanded to the point where it has established itself in recent decades as the most widely grown fine table grape in Brazil (Leão, 2000).

The plants should be grown under a long pruning regime, with poles of 8 to 12 buds. This cultivar has vigorous plants with a long cycle of around 150 days in the north-west of São Paulo and around 180 days in São Paulo, with an average yield of 30 tonnes/ha and little resistance to pests and diseases (Pommer et al, 2003). In the

Vale do Submédio São Francisco, it has a phenological cycle of around 120 days and an average yield of 30 tonnes/ha/year, which can reach up to 50 tonnes/ha/year in well-managed vineyards (Albuquerque, 1996).

The leaves are medium to large, pentalobate and orbicular. Large bunches, which can be very large, with an average weight of 450g, conical cylinders, sometimes winged, compact and therefore require intense thinning. The berries are large (8 to 12 g) and can reach more than 23 mm in diameter. They are oval, green or yellowish-green in colour, have a fleshy consistency, a neutral, slightly muscat flavour and adhere well to the pedicel. It is highly resistant to transport and storage (Leão, 2000).

The main problems with cv. Itália are related to its sensitivity to diseases, especially downy mildew and powdery mildew, and the need for intensive labour to carry out cultural treatments, especially thinning the berries, which cause losses and increase the cost of production. This has led many producers to look for other cultivars in order to diversify production (Tessmann et al, 2004).

The formation of the fertile bud is the consequence of the differentiation of the undifferentiated primordium into the reproductive primordium. This differentiation occurs in three well-defined stages (Mullins et al., 2000): formation of the undifferentiated primordia, which are protuberances of meristematic tissue from the apices of the latent buds. These tissues then develop into inflorescence primordia, tendril primordia, or shoot primordia, depending on various factors that influence the formation of the latent bud (Shikhamany, 1999).

Future plant structures are produced in the apical meristems. Some stem meristematic apices undergo major physiological changes, starting with the floral induction process, and transform from a vegetative to a reproductive form, the floral primordium or inflorescence. This process continues until the buds enter dormancy, induced by environmental factors such as temperature. During dormancy there are no visible changes in the size or differentiation of the primordium. The buds go through the dormant period already having the primordia of all the organs (leaves, tendrils and inflorescences) in an embryonic stage. In the following growing season, the bud sprouts and the branch starts growing again, completing a new differentiation process (Vieira, 1999).

The number of bunches is one of the main components of productivity and is determined by pruning and bud fertility, which is a quantitative measure of the plant's potential to produce fruit. The grapevine's performance in terms of budbreak and bud fertility can vary depending on the time of pruning and the grapevine cultivar (Leão e Silva, 2003), as well as the rootstock used (Pauletto et al, 2001 a).

Without pruning, the vine's productivity is not constant and the bunches are small and of poor quality. By limiting the number and length of branches, pruning provides a rational balance between vigour and production, regulating the quantity and quality of the grapes produced (Miele & Mandelli, 2004).

Knowing the position of the fertile buds for 'Itália' is of fundamental importance in defining the type of pruning to be used, as well as being a subsidy for establishing more appropriate pruning techniques that result in increased productivity. There is little information on the performance of this cultivar in tropical climate conditions.

Among the various practices used to manage the crop is green pruning, which includes: weeding, trimming, eliminating tendrils, eliminating grandchildren, thinning bunches, manual and/or chemical thinning and defoliation.

Defoliation consists of removing the leaves that cover the bunches, eliminating a maximum of one or two leaves per shoot, with the aim of balancing the leaf area/number of fruits ratio by improving ventilation and insolation inside the vineyard, thus achieving greater efficiency in controlling fungal diseases, especially in vigorous vineyards. This operation must be carried out with great care, as excessive defoliation can cause a lot of damage due to the lower accumulation of sugars in the fruit and incomplete ripening of the branches, as well as the occurrence of scald or "sunburn" on the berries. In vineyards where leaves overlap, more intense defoliation is necessary, eliminating all the leaves that are not exposed to sunlight (Silva & Correia, 2004).

The results presented below were extracted taking into account the influence of manual defoliation on bunch production and quality, as well as on the levels of carbohydrates contained in the leaves of the cv. Itália grapevine.

SUMMARY

The aim of this study was to assess the influence of manual defoliation on the quality of the bunches and the levels of carbohydrates contained in the leaves of the cv. Itália grapevine. The experimental design used was a randomised block design with a factorial scheme of 4 (5, 10, 15 and 20 leaves left per production branch) x 4 (phenological stages: 60, 72, 84 and 106 days after pruning) and nine replications, with one plant per experimental unit. At 60, 72 (berry growth), 84 (berry softening) and 106 (pre-harvest) days, photosynthesis, stomatal conductance and transpiration were assessed using a portable photosynthesis meter, IRGA, model LI-6200 (LI-COR) and the relative chlorophyll index (CRI) (SPAD unit), obtained using a SPAD-502 chlorophyll meter. The data for the variables were determined on healthy, fully expanded leaves exposed to the sun. The leaves were subjected to quantitative analyses of proteins, amino acids, starch, total sugars, reducing sugars and sucrose. The study also considered the following variables: number of shoots, yield per plant (kg), mass of fresh matter (g), diameter (mm), length (mm) and average volume (ml) of berries; soluble solids (°Brix), titratable acidity (g.tartaric acid/100 ml of pulp) and the soluble solids (SS)/titratable acidity (AT) ratio. For the physico-chemical evaluations of the berries, two bunches were collected per plant, from which all the berries were removed, followed by a random sample of 100 berries. The results obtained showed no significant differences between the treatments for the variables photosynthesis, stomatal conductance and transpiration, analysed during the growth and development phases of the berries, indicating that the removal of the leaves did not affect the gas exchange of the leaves opposite the bunch of cv. Itália vines, and that the content of reducing sugars in the leaves was lower during the initial phases of fruit development, showing an increase during the softening phase of the berries and in the pre-harvest phase. Sucrose was highest in the leaves during the berry softening and pre-harvest stages. The 15 leaves per branch treatment had a high starch content in the initial berry softening phase and a low starch content in the early fruit ripening phase. Higher concentrations of amino acids and proteins were observed in the softening and pre-harvest phases. The intensity of defoliation influenced the leaf content of the macromolecules analysed. Defoliation intensities did not interfere with the variables analysed, which generally showed positive correlations, except for the treatment with 10 leaves, whose associations of the variables number of shoots x berry diameter, production per plant x berry length, showed negative correlations.

Index Terms: Vitis vinifera, photosynthesis, carbohydrates

DEFOLIATION INTENSITIES ON THE PRODUCTION AND QUALITY OF GRAPEVINE BUNCHES CV. ITÁLIA IN THE SUB-MEDIUM SÃO FRANCISCO VALLEY

Table 1 shows that there were statistical differences between potential and actual bud fertility rates before pruning, reaching 51.53% and 39.66% respectively, i.e. 11.87% of buds considered fertile by microscopic analysis of cv. Itália buds were not expressed after budbreak. The number of inflorescence primordia that develop in the buds is influenced by varietal characteristics, among other factors. Ribeiro et al. (2008) found that the average rate of expression of actual/potential fertility of the Superior Seedless cv. grown in the Vale do Submédio São Francisco was 59.52%.

Table 1. Potential (PF) and Actual (FR) bud fertility of the cv. Itália grapevine. Petrolina-PE, 2010.

Characters	cv. Italia
FP (%)	51,53 a
FR (%)	39,67 b
F-test	4,97 *
CV (%)	31,957

Averages followed by different letters differ according to the Tukey test at 5% probability.

Differences in the percentage of fertile buds were observed in 14 grapevine cultivars studied, with Muscat of Alexandria, Centennial Seedless and Italia having the highest percentage values, while Redglobe and Superior Seedless had the lowest. Shikhamany (1999) relates the problems of low bud fertility to the high vegetative vigour of the plants, especially in seedless table grape cultivars. There is also the influence of environmental conditions, which may be related to the existence of genes that alter plant metabolism, such as the production of gibberellins (Boss et al., 2003).

The different levels of defoliation did not influence the response of the variables number of shoots, production per plant, diameter, length, volume, soluble solids, titratable acidity and the SS/AT ratio of the berries of the cv. Itália grapevine (Table 2).

Table 2. Summary of the analysis of variance (F test) and coefficients of variation (CV) for the variables number of shoots, yield per plant (kg), diameter (mm), length (mm), volume (ml), soluble solids (°Brix), titratable acidity (g.tartaric acid/100g of pulp) and SS/AT ratio of berries from the cv. Itália grapevine, managed with 5, 10, 15 and 20 leaves per production branch. Petrolina-PE, 2010.

Characters	F-test	CV (%)
Number of shoots	0.573 ns	27,80

Production per plant	2,363 ns	27,51
Berry diameter	0.204 ns	3,42
Berry length	0.095 ns	2,97
Volume of berries	0.141 ns	11,72
Soluble Solids (SS)	1.015 ns	3,32
Titratable acidity (TA)	2,089 ns	8,47
SS/AT ratio	1.736 ns	10,50

ns = not significant at 5% probability

Positive correlations were observed between the DB variable and VB, CB, VB and CB, respectively, i.e. there was a close association between the characteristics related to the growth (diameter, volume and length) of cv. Itália berries, for all the defoliation intensities (Table 3). Positive correlations between fruit dimensions are to be expected regardless of the defoliation intensities applied, confirming only that the treatments were correctly applied.

Table 3. Correlation coefficient between the variables, number of shoots, production/plant (Kg), berry diameter (mm), berry length (mm), berry volume (ml), soluble solids (°Brix) and titratable acidity (g.tartaric acid/100g of pulp), of cv. Itália managed with 5, 10, 15 and 20 leaves per production branch. Petrolina-PE, 2010.

Characters	5 sheets	10 sheets	15 sheets	20 sheets
Berry diameter x Berry volume	0,9364**	-	-	-
Number of shoots x Berry diameter	-	-0,7696*	-	-
Yield/plant x Berry length	-	-0,7153*	-	-
Berry diameter x Berry length berries	-	0,6850*	-	-
Production/plant x Berry diameter	-	-	0,6974*	-
Berry diameter x Berry volume	-	-	0,6934*	-
Berry length x Titratable acidity	-	-	0,9161**	-
Berry diameter x Berry length berries	-	-	-	0,8476**

** significant at 1% probability ($\alpha < 0.01$) * significant at 5% probability ($\alpha < 0.05$)

As for plant yield, 10 and 15 leaves were the only defoliation indices that were correlated with this characteristic, indicating that berry length and diameter have negative and positive correlations, respectively, with yields, with the 15-leaf index increasing the yield (kg/plant) of cv. Itália. Scarpare Filho et al. (2010) observed lower yields of the Niágara Rosada cv. in kg/plant, due to the reduction in leaf area caused by the incidence of fungal diseases, with the diseased plants showing only 55% of the yield obtained by the treatment without defoliation and with phytosanitary control.

In plants that had 10 leaves per branch, there was a negative correlation between the variables number of shoots and berry diameter. Manipulation of sources and drains can affect the rate of photosynthesis in grapevines. Removing bunches can reduce photosynthetic activity, while partially removing leaves can increase it (Kliewer, 1990). This was not the case in this study, indicating that there must have been a translocation of reserve substances for the development of shoots to the detriment of the diameter of the berries and, consequently, the 10 leaves kept per branch did not produce enough photosynthates for the formation of shoots and berries, which are characterised by being strong metabolic drains for the vine.

For the 15-leaf intensity, there was a close positive association between berry length and titratable acidity, with a high correlation coefficient (0.9161) and significant at 1% probability. This result is not corroborated by Felippeto (2008), who reports that the tartaric acid content in grape must is approximately 15 g L^{-1} during the growing season.

at the fruit formation stage, decreasing to 6 g L^{-1} to 7 g L^{-1} at the ripening stage, mainly due to its dissolution in the must as a result of the size of the berry. In the same study, the author found lower acidity with the 50% and 75% defoliation treatments after defoliation intensities of 0, 25%, 50% and 75% on Cabernet Sauvignon. According to Chitarra and Chitarra (2005) organic acids are translocated from the leaves to the fruit and are used as a respiratory substrate and in protein synthesis, increasing progressively as the fruit grows and showing a reduced concentration as it ripens, when its growth ceases; however, the authors also point out that the concentration of organic acids does not decline in all types of fruit during ripening, so there are exceptions, and that fruit from partially defoliated plants are less acidic than those from defoliated plants. On the other hand, the phenological cycle of cv. Itália in semi-arid tropical conditions is approximately 120 days (pruning to harvest), and in order to obtain a better flavour intensity it must be harvested with at least 16 °Brix (Pommer et al., 2003). In the north of Rio de Janeiro, Murakami et al (2002) found phenological cycles of 138 to 157 days for cv. Itália, for pruning carried out in April and July, respectively. In this work, it was observed that at 115 days after pruning, the berries had an average soluble solids content of 13.733 ± 0.519 °Brix and a titratable acidity of 1.028 ± 0.081 g.tartaric acid/100 g of pulp (Table 4), which is therefore the cause of the high titratable acidity found in the berries.

Table 4 Means and standard deviation of soluble solids (SS, in °Brix), titratable acidity (AT, in g.tartaric acid/100 g of pulp), SS/AT ratio, diameter (mm), length (mm) and volume (ml) of grapevine cv. Itália berries, as a function of defoliation intensities (5, 10, 15 and 20 leaves per production pole). Petrolina-PE. 2010

Characters	5 sheets	10 sheets	15 sheets	20 sheets
Soluble solids	13,855 ± 0,269	13,488 ± 0,735	13,733 ± 0,519	13,722 ± 0,423
Titratable acidity	1,096 ± 0,044	1,119 ± 0,102	1,028 ± 0,081	1,040 ± 0,097
SS/AT ratio	12,651 ± 0,517	12,170 ± 1,548	13,461 ± 1,479	13,291 ± 1,319
Diameter	21,432 ± 0,941	21,596 ± 0,491	21,548 ± 0,855	21,701 ± 0,656
Length	23,960 ± 0,753	23,837 ± 0,705	23,942 ± 0,627	23,815 ± 0,739
Volume	6,102 ± 0,829	6,237 ± 0,547	6,099 ± 0,727	6,237 ± 0,758

Thus, the intensities of defoliation did not affect production or the physical and chemical characteristics of the fruit.

PHOTOSYNTHESIS, TRANSPIRATION AND STOMATAL CONDUCTANCE OF CV. ITÁLIA IN THE SUB-MEDIUM SÃO FRANCISCO VALLEY

ABSTRACT - The aim of this study was to evaluate the effect of defoliation on the photosynthetic activity of cv. Itália grapevine leaves. The experiment was carried out in a commercial vineyard, conducted in a trellis system, located in the Senador Nilo Coelho Irrigated Perimeter, in Petrolina - PE. The experimental design used was a randomised block design with a factorial scheme of 4 (5, 10, 15, 20 leaves left on the production branch) X 4 (phenological phases at 60, 72, 84, 106 days after pruning) and six replications, giving a total of 96 readings. The leaves most exposed to sunlight, fully expanded, opposite the bunch, without signs of senescence and healthy were selected. Photosynthesis, stomatal conductance and transpiration were assessed using a portable IRGA. These variables were studied on healthy, fully expanded leaves exposed to the sun. The results obtained showed no significant differences between the treatments in relation to the variables studied during the growth and development phases of the berries, indicating that the removal of the leaves did not affect the gas exchange of the leaves opposite the cluster of the cv. Itália grapevine.

Index Terms: Vitis vinifera L., defoliation, gas exchange.

The photosynthetic rate (Table 1) at the different stages of plant development shows that there was no significant difference between the berry growth and pre-harvest stages. As photosynthetic rates vary depending on environmental conditions, Table 2 shows that the meteorological data for December 2009 and January 2010 showed average monthly temperatures (27.4°C and 27.1), average monthly minimum temperatures (23°C and 22°C), average maximum temperatures (33.7°C and 33.8°C) and relative humidity (60% and 61%), which were very close, respectively.

Table 1- Photosynthetic rate (A); stomatal conductance (gs); (CINT) and transpiration (E) in leaves of grapevines of the Italia cultivar at different stages of berry growth. Petrolina, 2010

Treatments	**A (μmol m^{-2} S^{-1})**	CINT	gs (mol m^{-2} S^{-1})	E (mmol)
60 DAP	10,981 a	261,060 a	0,806 a	7,453 ab
72 DAP	8,869 a	233,317 b	0,381 b	4,642 c
84 DAP	10,944 a	262,775 a	0,748 a	8,470 a
106 DAP	10,237 a	273,765 a	0,723 a	7,169 b
F-test	2.671 ns	8,018 **	7,147 **	17,779 **
5 sheets	10,019 a	255,801 a	0,651 a	6,542 a
10 sheets	10,472 a	252,826 a	0,676 a	6,746 a
15 sheets	9,836 a	263,951 a	0,615 a	6,920 a
20 sheets	10,705 a	258,339 a	0,716 a	7,525 a
F-test	0.439 ns	0.602 ns	0.349 ns	1.204 ns
Interaction A X B	1.454 ns	0.996 ns	0.892 ns	0.578 ns
Treatments	1.495 ns	2,321 **	2,034 *	4,143 **
CV(%)	28,848	11,559	52,986	27,251

** significant at 1% probability level (p < ,01)
* significant at 5% probability level (,01 =< p < ,05)
ns not significant (p >= ,05)

Table 2- Monthly rainfall (PPM), average monthly temperature (TM), average monthly relative humidity (UR), average global solar radiation (RS), average monthly minimum temperature (TMM), average evaporation (EV), average monthly insolation (IS), average maximum temperature (TMA), average monthly wind speed at a height of 2.0m (VV), in the region during the experiment. Bebedouro Agrometeorological Station. Petrolina-PE. Period 2009-2010.

Climate data	PM (mm)	TM (°C)	RH (%)	RS (ly/day)	TMM (°C)	EV (mm)	IS (hours)	TMA (°C)	VV (Km/day)
Dec 2009	49,0	27,4	60	347,8	23,0	7,3	5,4	33,7	149,8
Jan 2010	8,9	27,1	61	360,5	22,4	7,8	6,9	33,8	154,5

According to Lima Filho et al, 2009, the temperature range considered optimal for grapevine leaves to express their maximum photosynthetic capacity is between 25°C and 30°C. In this temperature range, biochemical reactions take place, the catalysts of which, enzymes, are dependent on temperature to express their maximum activity. The photosynthetic rate is compromised mainly by the reduction in stomatal conductance (Chaves & Oliveira, 2004).

At 72 days after pruning, conductance (Table 1) was 0.381mol m-2 s-1, lower than in the other phases. The reduction in conductance may have been influenced by various environmental factors, in agreement with Centrito (2003), who states that a reduction in conductance is one of the first responses to environmental stress. However, in the other phases of the plant there was an increase in stomatal conductance, indicating a recovery of the plant, as there were no biochemical limitations on the photosynthetic rate. This is the physiological mechanism that vascular terrestrial plants have for controlling transpiration (Messinger et al., 2006). According to Naves-Barbiero et al. (2000), the leaf epidermis is surrounded by a cuticle that is relatively impermeable to both water vapour and carbon dioxide and contains a variable number of stomata, whose response, through the regulation of stomatal conductance, controls leaf transpiration and is influenced by light, atmospheric carbon dioxide concentration, humidity and temperature. According to these authors, stomatal conductance is proportional to transpiration, net photosynthesis and leaf water potential.

Table 1 shows that the greatest transpiration was observed at 60, 84 and 106 days after pruning. This is beneficial as this factor is directly correlated with the photosynthetic rate. The decrease at 72 days indicates that the plant has undergone some stress caused by climatic factors, where in Table 3 photosynthetically active radiation (PAR) was higher than 1700 |imoLm'^{2}.S'^{1}, temperature higher than 34°C and relative humidity varying between 43% and 45%, observed on the leaves inside the IRGA chamber and/or excessive application of some chemical product (Vidal, 1997).

The transpiration of most plant species, including grapevines, is determined by climatic demands related to solar radiation, physiological mechanisms related to stomatal responses to environmental factors, leaf area index and soil water availability (Taiz & Zeiger, 2009).

Stomata play an important role in the thermal control of the leaf. Under conditions of transpiration and in relation to air temperature, leaf temperature can be reduced by up to 8°C (Wilkinson, 2004).

It is worth noting that the variables photosynthetic rate, conductance and transpiration showed no interaction or significant difference with the number of leaves left on the production branch in the cv. Itália vine.

Table 3- Photosynthetically active radiation (PAR) (μmol.m^{-2}.s^{-1}); Temperature (°C); Leaf temperature (°C) and

Relative humidity (%) observed on the leaves and their effect on the photosynthesis of the cv. Itália grapevine at the different defoliation rates.

Different defoliation rates	Different phenological phases	Radiation µmol.m^{-2}.s^{-1})	Leaf temperature (°C)	Relative humidity (%)
5 sheets	60 D AP	1283,21	34,85	54,27
	72 DAP	1537,15	34,96	45,96
	84 DAP	1238,49	36,60	49,93
	106 DAP	851,62	32,17	48,21
10 sheets	60 DAP	1501,38	36,69	50,55
	72 DAP	1556,12	36,54	46,25
	84 DAP	1581,85	37,53	47,96
	106 DAP	1459,44	35,08	47,63
15 sheets	60 DAP	1461,44	35,93	49,43
	72 DAP	880,18	34,87	44,58
	84 DAP	1337,84	38,23	49,73
	106 DAP	1084,13	34,85	50,70
20 sheets	60 DAP	1418,55	35,69	50,32
	72 DAP	1788,00	38,72	43,30
	84 DAP	1347,03	38,60	52,42
	106 DAP	1216,65	34,16	50,37

Therefore, leaf removal did not alter photosynthetic activity, stomatal conductance and transpiration in the cv. Itália grapevine. It can be used in cultural management.

DIFFERENT LEVELS OF DEFOLIATION AND CARBOHYDRATE AND PROTEIN CONTENT IN VINE LEAVES CV. ITALY

Abstract - The aim of this study was to evaluate the influence of manual defoliation and phenological phase on the levels of carbohydrates, proteins and chlorophyll content contained in the leaves of the cv. Itália grapevine. The experimental design used was randomised blocks in a 4 (5,10,15, 20 leaves left on the production branch) X 4 (phenological phases at 60, 72, 84, 106 days after pruning) factorial design with three replications. The leaves most exposed to sunlight, fully expanded, opposite the bunch, without signs of senescence and healthy were selected. One leaf per treatment was taken, totalling 96 samples. The leaves were subjected to quantitative analyses of proteins, amino acids, starch, total sugars, reducing sugars and sucrose. The starch content was high at the initial stage of berry softening and with 15 leaves left on the production branch, and low at the beginning of the fruit ripening stage. The accumulation of carbohydrates in the leaves was influenced by the phenological stage of the vine and the intensity of defoliation influenced the leaf contents of the macromolecules analysed.

Index Terms: Vitis vinifera L., leaf number, macromolecules

It can be seen in Figures 1a that the content of reducing sugars (RS) in the leaves was lowest during the early stages of fruit development, showing an increase during the berry growth phase and in the pre-harvest phase, in agreement with Takayanagi & Yokotsu (1997), who state that there is a large accumulation of glucose and fructose in the berries during ripening due to the large transport of photoassimilates needed to form them.

Yokotsuka (1997), who state that there is a large accumulation of glucose and fructose in the berries during ripening due to the large transport of photoassimilates necessary for the formation of structural sugars needed during berry growth, which was observed in this study. Sucrose (Figure 1a) is present in greater quantities in the leaves, in the growth phase and in the pre-harvest phase. According to Hunter et al. (1994), sucrose levels have an inverse behaviour when compared to RA. According to Lawlor and Cornic (2002), when sucrose synthesis is limited, translocation can reduce the concentration of this molecule in the leaves. In figure 1c, it can be seen that at the beginning of berry growth (strong drain) the leaf content of total soluble sugars (TSS) is lower, increasing during ripening and the highest leaf content of TSS was found at pre-harvest. The leaf starch content was high at the start of berry growth, due to the high photosynthetic rate (Fig 3, chap II). At this stage, in general, there is a low leaf sugar status due to the strength of the drain imposed by the growing berries, which improves photosynthesis, reserve mobilisation and export. On the other hand, abundant sugar promotes the growth and storage of carbohydrates in the drains and, when the rate of photosynthesis is high, there is an accumulation of AST and starch in the leaves (Taiz and Zeiger, 2009).

In the pre-harvest phase, there was a low leaf starch content and an increasing leaf AST content. This was due to the greater consumption of sugars by the drains in relation to their production by the sources. At stages when there is not much transport of sugars from the photosynthesising leaves to the ripening fruit, such as the final stages of the production cycle, starch tends to accumulate in the leaves, regulating photosynthetic activity (Pimentel, 1998). Dantas et al (2007), studying the Syrah cultivar, found a starch content in the youngest leaves of 1.85mg g^{-1}MF, which they considered to be low. In the leaf after the bunch, they found 12.16 mg g^{-1}MF, considered to be the peak in starch content. During ripening, when photosynthesis was lowest (Figure 3, Chapter II), there was a reduction in leaf starch content. Photosynthesis peaked at 60 days after pruning and decreased during pre-harvest at 106 days after pruning, at which stage the leaves can be considered a weak drain (Figure 1c). During ripening, the bunches are the main drains (Kliewer, 1981). Therefore, all the sugars produced by the leaves are translocated to the berries in priority to the apical meristems, accumulating total sugars and starch in the leaves close to the bunch. The sugars in the leaves may reflect the metabolic state of photosynthesis or also the ability to translocate these sugars to reserve tissues. No statistical difference was found in the leaf starch and AST content at the different defoliation intensities (Figure 1d).

Amino acid and protein content (Figure 1e) were higher in the growth and pre-harvest phases. Amino acids are involved in a large part of primary and secondary metabolism, leading to the synthesis of various compounds that influence fruit production and quality. The amino acids asparagine and glutamate link the plant's two most important metabolic cycles, the carbon cycle and the nitrogen cycle, influencing the production of both sugars and proteins. Glycine is an amino acid that inhibits the apparent photorespiration carried out by C3 plants, such as grapevines, leading to greater photosynthesis efficiency, and thus higher sugar content and productivity (Taiz & Zeiger, 2009). The production of correctly structured proteins and their maintenance are important for efficient cell function (Lawlor, 2002). As amino acids are precursors to protein, an increase in amino acids consequently leads to an increase in protein.

Plants with 15 leaves had a higher photosynthetic rate, starch (Figure 1d), reducing sugars (Figure 1b), and a lot of sucrose transport (Figure 1b) and a lower amino acid content (Figure 1f). For this amount of leaf, the plant showed greater leaf efficiency (greater transport and accumulation).

Sucrose and reducing sugars (glucose and fructose) are synthesised in the leaves and translocated during the day and night to all parts of the plant via the phloem. Starch is not translocated. Therefore, the translocation of sugar supplies may originate from photosynthesis during the day and starch consumption at night (Galdiano, 2008). According to Borba (2005), when a plant enters active growth, carbohydrate metabolism becomes more intense and the increase in sucrose occurs at the expense of carbohydrate production through photosynthesis. The increase in photosynthetic activity and, consequently, the accumulation of sugars in expanding leaves is due to an increase in chlorophyll content per unit leaf area, an increase in the activity of carboxylative enzymes and a decrease in stomatal resistance (Kliewer, 1981).

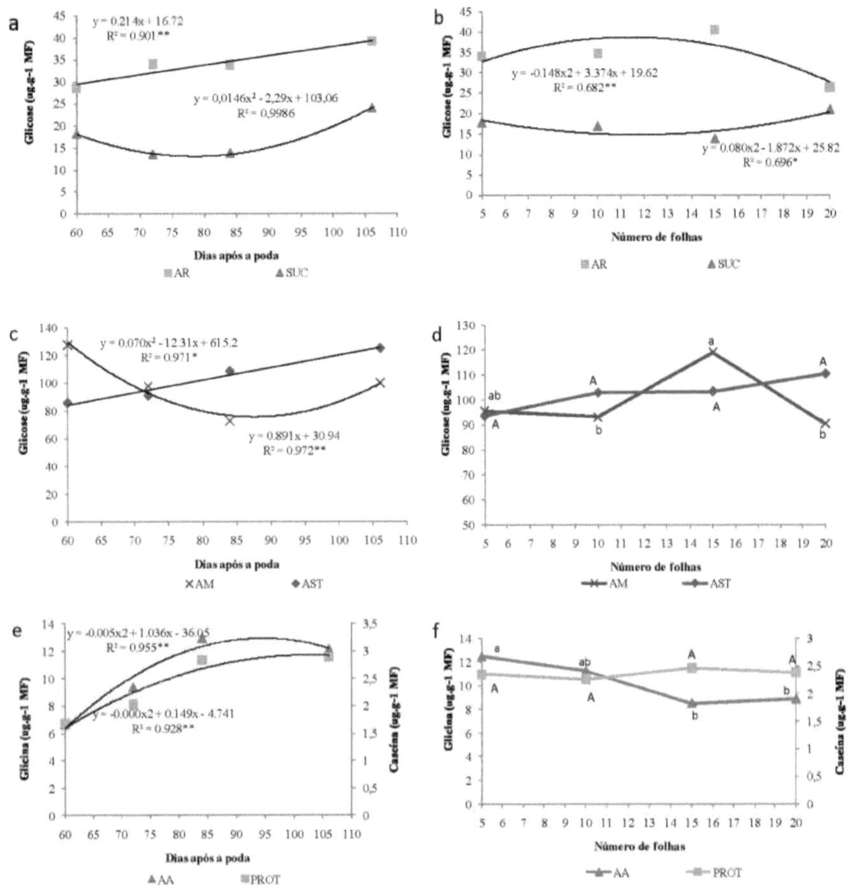

Figura 1. Leaf contents of reducing sugars and sucrose (a, b); starch and total sugars (c, d); amino acids and protein (e, f) (mg.g^{-1} of fresh material) in grapevine cultivar Italia at different phenological stages and with different amounts of leaves left on the production branch, Petrolina- PE, 2009 to 2010.

Figure 2 shows that the chlorophyll content increased during the physiological phases, with its highest concentration being 42.15 SPAD at 106 days after pruning, in the pre-harvest phase, in line with the results for proteins and amino acids in the leaves. According to Taiz & Zeiger (2009), the chlorophyll content in leaves is influenced by various biotic and abiotic factors and is directly related to the plant's potential for photosynthetic activity. In general, the increase in chlorophyll concentration with the reduction in luminosity increased the capacity to absorb light of wavelengths other than the peaks of photosynthesis, such as light in the green range. In general, chlorophyll and carotenoids tend to increase in concentration as light intensity decreases (Ferraz & Silva, 2001; Fontes & Silva, 2001).

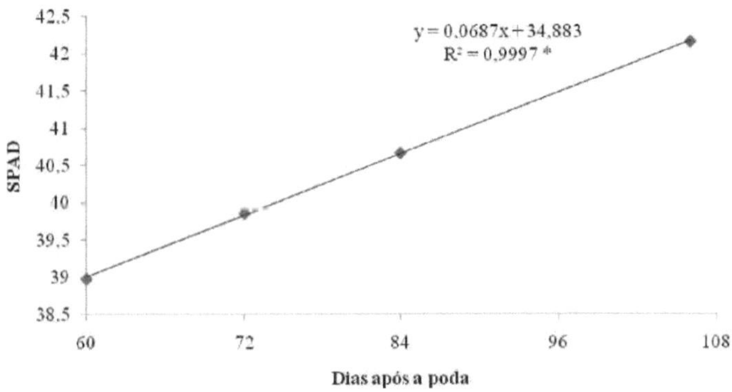

Figura 2. Foliar chlorophyll content (SPAD) in grapevine cultivar Itália at different stages of berry growth, Petrolina-PE, 2009 to 2010.

Thus, the intensity of defoliation influenced the leaf content of the macromolecules analysed.

CHAPTER 3

PLANT REGULATORS AND FOLIAGE FERTILISER APPLIED IN THE GROWTH PHASE OF 'SWEET SUNSHINE' VINE BERRIES IN THE I° SEMESTER OF 2013

The region of the São Francisco Valley, whose centre cities are Petrolina in Pernambuco and Juazeiro in Bahia, is a pioneer in the production of grapes and wines under tropical conditions in Brazil. The production of grapes in these conditions depends on vine management technology, which has been developed and adjusted over the last 50 years, reaching a high level of technology today. The region's climatic conditions make it possible to carry out 2.5 vegetative cycles per year and to programme the harvest time for any day of the year, which makes it possible to have a constant supply of table grapes on the market and to process large volumes of grapes throughout the year in small wineries (PROTAS & CAMARGO, 2011).

In recent years, there has been a considerable increase in consumer demand for seedless grapes. Currently, the Thompson Seedless, Superior Seedless and Crimson Seedless cultivars are the most widely grown seedless grapes in the São Francisco Valley. However, producers are investing in new seedless varieties, including the Sweet Sunshine cv, of American origin (USA).

This and other table grape varieties grown in the São Francisco Valley sub-medium were the result of the introduction of varieties from other countries, varieties that were developed for other environments in which the climatic conditions are very different; therefore, although they stand out for their commercial attributes, they require adjustments in the management of the use of pesticides, plant regulators, organic and mineral fertilisation.

The Sweet Sunshine cultivar, in addition to having excellent commercial characteristics, is rain-tolerant, but needs to be adjusted in terms of cultural management, as it has a high degranulation index, which is related to the adherence of the berry to the pedicel, which reflects the strength with which the berry is held to the pedicel in the rachis. This cultivar is also sensitive to the use of gibberellins. Therefore, the application of a mixture of plant regulators (IBA+GA3+Kt) combined with foliar fertiliser (N + **P2O5**+ S+ Co + Mo) can promote synergism in the plant's hormonal balance and the foliar fertiliser, as well as promoting nutritional balance, acts on the synthesis of these hormones.

In addition, increasingly intensive viticultural exploitation has led to a profound imbalance in the vines, altering their metabolism and, consequently, the hormonal balance of the plants (BOTELHO et al., 2006). This has led to higher production costs, making grape growing in the sub-middle São Francisco Valley region less competitive.

Therefore, the phenological characterisation of the Sweet Sunshine grapevine cultivar for the conditions of the sub-middle São Francisco Valley is necessary in order to identify the different stages of the plant's development, such as budding, flowering, fruiting, fruit ripening and harvesting. By defining these stages of development, it is possible to define the managements commonly used in the vineyard and try to increase the plant's productivity.

The aim of this study was therefore to assess the effect of foliar application of plant regulators and foliar fertiliser on the physiological and productive behaviour of Sweet Sunshine grapevines, as well as on reducing the rate of degeneration and increasing fruit production and quality.

SUMMARY

The aim of this study was to evaluate the effects of plant regulators and foliar fertiliser on the physiological and productive behaviour and the reduction in the degradation index of the 'Sweet Sunshine' grapevine in the first half of 2013. A single application of a mixture of plant regulators combined with foliar fertiliser was applied during the berry growth phase, with the following treatments: Witness; Plant regulators 1.6 L ha^{-1} + foliar fertiliser 1.2 L ha^{-1}; Plant regulators 1.6 L ha^{-1}+ foliar fertiliser 1.5 L ha^{-1}; Plant regulators 1.8 L $ha^{(-1)}$+ foliar fertiliser 1.2 L ha^{-1} and Plant regulators 1.8 L ha^{-1} + foliar fertiliser 1.5 L $ha^{(-1})$. The data was subjected to analyses of covariance. The effects of applying plant regulators and foliar fertiliser were assessed by physiological and biochemical analyses of the leaves, and biometric and chemical analyses of the fruit. It was found that the physiological parameters promoted changes in plant characteristics between the two periods, as the behaviour of the gas exchange measurement curves was quite different. The characteristics of the fruit show that this cultivar has productive and commercial potential. For the best response to reduce berry degranulation in grapevine bunches, we recommend a mixture of plant regulators (IBA + GA3+ Kt) at 2.0 L ha^{-1} and 1.6 L ha^{-1} applied alone and/or combined with foliar fertiliser at 1.5 L ha^{-1}, in the pre-flowering, flowering, fruiting and softening phases, in the 1st and 2nd half of the year, respectively, under the conditions of the São Francisco sub-medium region.

Keywords: **Vitis vinifera** L., gas exchange, antioxidant enzymes, photosynthetic pigments, bohydrates

Table 1 shows that there were statistical differences between the treatments for **CO2** assimilation rate (**A**, µmol **CO2** m^{-2} $s^{-1)}$, stomatal conductance (*gs*, mol m^{-2} s^{-1}), transpiration rate (**E, mmol water vapour m**$^{-2}$ $s^{-1)}$, internal **CO2** concentration (*Ci/Ca)* and carboxylation efficiency (*A/Ci)*. While for days after application (DAA) the differences were seen in the variables A; *g;E;* water use efficiency *(EUA,* µmo , **CO**$^{(}_{2)}$ (mmol **H2O**$^{)(-1)}$); *A/Ci* and leaf temperature (TF, °C), while in the interaction only *CiCa* and *A/Ci* were significant, in 'Sweet Sunshine' vines in the first half of 2013.

Table 1. CO2 assimilation rate (A, µmol **CO2** m^{-2} $s^{-1)}$, stomatal conductance *(gs,* mol m^{-2} s^{-1}), transpiration rate (E, mmol water vapour m^{-2} $s^{-1)}$, internal $CO_{(2)}$ concentration *(Ci/Ca),* water use efficiency *(WUE,* pmol **CO**$_{(2)}$) (µmol $H_2O)^{-1}$); carboxylation efficiency (A/Ci) and leaf temperature (TF, °C) at 5, 13 and 20 days after application of the mixture of plant regulators combined with foliar fertiliser in the berry growth phase (10-12 mm) of the 'Sweet Sunshine' grapevine, in the submedium of the São Francisco Valley, Petrolina-PE, 2013.

Dependent Variables	Independent Variables	Prob. F test
A (pmol $_{CO2}$ m^{-2} s^{-1})	Treatment	0,0041

	DAA	<0,0001
	DAA* Trat	0,0033
gs (mol m^{-2} s^{-1})	Treatment	0,0073
	DAA	0,0120
	DAA* Trat	0,2399
E (mmol water vapour m^{-2} s^{-1})	Treatment	0,0254
	DAA	<0,0001
	DAA* Trat	0,5994
Ci/Ca	Treatment	0,0058
	DAA	0,1523
	DAA* Trat	0,0003
USA (µmol CO2 (mmol $_{H2O}$)$^{(-1)}$)	Treatment	0,4482
	DAA	<0,0001
	DAA* Trat	0,3515
A/Ci	Treatment	0,0130
	DAA	<0,0001
	DAA* Trat	0,0027
TF (°C)	Treatment	0,0571
	DAA	<0,0001
	DAA* Trat	0,8193

Figure 30 shows the significant effect of the treatments on the CO_2 assimilation rate (A, µmol CO_2 m^{-2} s^{-1}), stomatal conductance (gs, mol m^{-2} $s^{(-1)}$)) and transpiration rate (E, mmol water vapour m^{-2} $s^{(-1)}$)), with the mixture of RV 1.8 L $ha^{(-1)}$ + FF 1.2 L $ha^{(-1}$) showing the best results. This greater stomatal opening led to greater CO_2 diffusion into the leaf mesophyll, promoting a higher assimilation rate.

There was also a sharp drop in these variables throughout the grapevine cycle, reaching high values at 5 days after application (5DAA), favouring maximum absorption of photoassimilates, and regressing at 13 and 20 DAA, which may be related to the redistribution of photoassimilates in these phases. As for internal CO2 concentration (*Ci/Ca*), the best results were achieved at the dose of the mixture of RV 1.6 L ha^{-1} + FF 1.2 L ha^{-1} at 5 and 13 DAA, reducing at 20 DAA. This indicates the time taken for this treatment to act on the plant.

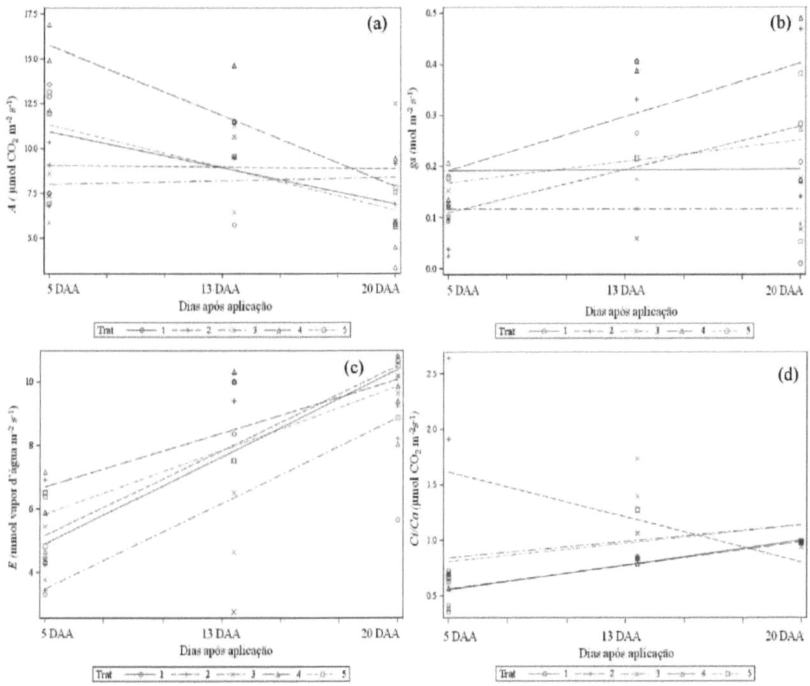

Figure 1. Covariance analysis for CO_2 assimilation rate (A, µmol $\mathbf{CO_2}$ m^{-2} $s^{-1)}$, stomatal conductance (gs, mol m^{-2} $s^{(-1)}$)), transpiration rate *(E,* mmol water vapour m^{-2} $s^{(-1)}$)), internal **CO2** concentration *(Ci/Ca,* µmol **CO2** $m^{(-2)}s^{-1}$), at 5, 13 and 20 days after application of the mixture of plant regulators combined with foliar fertiliser in the berry growth phase (10-12 mm) of the 'Sweet Sunshine' grapevine, in the sub-middle of the São Francisco Valley, Petrolina-PE, 2013.

DAA - Days after application; Trat. - Treatments; RV - Plant regulators; FF - Foliar fertiliser; T1- Witness; T2- RV 1.6 L ha^{-1} + FF 1.2 L ha^{-1} ; T3- RV 1.6 L ha^{-1} + FF 1.5 L ha^{-1} ; T4- RV 1, 8 L ha^{-1} + FF 1.2 L ha^{-1}; T5- RV 1, 8 L $ha^{(-1)}$+ FF 1.5 L $ha^{(-1)}$

Transpiration rates (E) followed those of stomatal conductance (gs). The increase in transpiration rate (E) as an effect of the increase in gs indicates that the treatments acted as a protective mechanism against water loss until 13 DAA, due to the climatic conditions of the semi-arid region. It can also be seen that high gs values induced reductions in A values (Fig. 30 a), water use efficiency (Fig. 31 a) and carboxylation efficiency (Fig. 31 b).

Analysing leaf temperature (FT) at 5, 13 and 20 DAA, the treatments with VR at 1.6 L ha^{-1} + 1.5 L ha^{-1} of FF and VR at 1.8 L $ha^{(-1)}$+ 1.5 L $ha^{(-1}$) of FF showed the highest values, differing statistically from the other treatments. This higher leaf temperature was not reflected in an increase or decrease in the **CO2** assimilation rate.

The results also show that the effect of the treatments was more evident at 5 and 13 DAA, while at 20 DAA, all the treatments showed a reduction in responses for most of the variables evaluated, indicating that plant regulators and minerals can be active in plants for at least 13 days and after this period, their activity decreases.

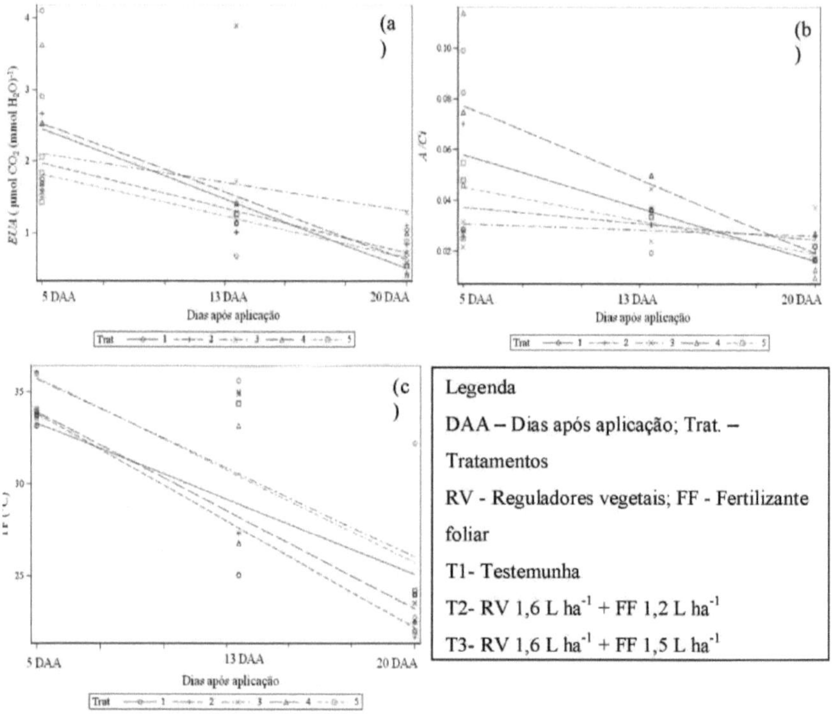

Figure 2. Covariance analyses for water use efficiency *(WUE,* **µmol CO_2** (mmol $H_2O)^{-1}$); carboxylation efficiency (*A/Ci*) and leaf temperature (TF, °C) at 5, 13 and 20 days after application of the mixture of plant regulators combined with foliar fertiliser at the berry growth stage (10-12 mm) of the 'Sweet Sunshine' grapevine, in the sub-middle São Francisco Valley, Petrolina-PE, 2013.

Table 2 shows that there was a statistical difference between treatments for reducing sugars, amino acids, superoxide dismutase, catalase and peroxidase, while for days after application (DAA) total soluble sugars, reducing sugars, amino acids, proteins, chlorophyll *b,* superoxide dismutase, catalase and peroxidase, and in the interaction total sugars, reducing sugars, sucrose, amino acids, superoxide dismutase, catalase and peroxidase.

Table 2. Foliar contents of total soluble sugars, reducing sugars, sucrose, amino acids, proteins, chlorophyll a and b µg g^{-1} of fresh matter (MF), superoxide dismutase, catalase and peroxidase Umg $protein^{-1}$, and Lipid peroxidation nmol g MF^{-1} at 5, 13 and 20 days after application of the mixture of plant regulators combined with foliar fertiliser in the berry growth phase (10-12 mm) of the 'Sweet Sunshine' grapevine, in the sub-middle of the São Francisco Valley, Petrolina-PE, 2013.

Dependent Variables	Independent Variables	Prob. F test
Total sugars (μg g^{-1} MF)	Treatment	0,1656
	DAA	0,7390
	DAA* Trat	0,0001
Reducing sugars (μg g^{-1} MF)	Treatment	<0,0001
	DAA	<0,0001
	DAA* Trat	<0,0001
Sucrose (μg g^{-1} MF)	Treatment	0,2839
	DAA	0,3358
	DAA* Trat	<0,0001
Amino acids (μg g^{-1}MF)	Treatment	<0,0001
	DAA	0,0001
	DAA* Trat	0,0009
Proteins (μg g^{-1} MF)	Treatment	0,5770
	DAA	0,0370
	DAA* Trat	0,7973
Chlorophyll a (μg g^{-1} MF)	Treatment	0,2200
	DAA	0,5282
	DAA* Trat	0,7801
Chlorophyll b (μg g^{-1} MF)	Treatment	0,4312
	DAA	0,0456
	DAA* Trat	0,6827
Superoxide dismutase (Umg $protein^{-1}$)	Treatment	<0,0001
	DAA	0,0202
	DAA* Trat	0,0002
Catalase (Umg $protein^{-1}$)	Treatment	<0,0001
	DAA	0,0302
	DAA* Trat	0,0001
Peroxidase (Umg $protein^{-1}$)	Treatment	<0,0001
	DAA	<0,0001
	DAA* Trat	<0,0001
Lipid peroxidation (nmol g MF^{-1})	Treatment	0,8465
	DAA	0,1818
	DAA* Trat	0,8441

Analysing the total sugar content (Figure 32a) showed a progressive increase in these sugars throughout the vine cycle, with the highest demand at 20 DAA, the berry growth and development phase, indicating that the treatments may have favoured the accumulation of photoassimilates in the leaves, since grapevine plants have a higher demand for photoassimilates in the fruit growth and development phases.

For the content of reducing sugars (RA), at 13 and 20 DAA, the treatments with plant regulators at 1.6 L ha^{-1} + 1.2 L ha^{-1} of foliar fertiliser and RV at 1.8 L ha^{-1} + 1.5 L $ha^{(-1)}$) of FF were those with the highest averages (Figure 3 b). Thus, there was a greater accumulation of reducing sugars, glucose and fructose, in the leaves in the berry softening phase, justifying the demand for carbohydrates in the different phenological phases of grapevine plants.

For sucrose (Figure 3c), at 5 DAA, the treatments with plant regulators associated with foliar fertiliser (1.6 L ha^{-1} + 1.5 L ha^{-1}; 1.8 L ha^{-1} + 1.2 L $ha^{(-1)}$); 1.8 L ha^{-1} + 1.5 L ha^{-1}, respectively) showed high levels, regressing at 13 and 20 DAA, making it possible to store reserves in organs such as branches and roots, this being the first reserve carbohydrate found in the leaves. At 20 DAA, the control plant and the plants treated with RV 1.6 L ha^{-1} + 1.2 L ha^{-1} FF showed high levels at 20 DAA, which was the opposite when compared to the results at 13 and 5 DAA, indicating that the control plant and this treatment were deficient in sucrose translocation, which resulted in a lower total bunch mass per plant (Figure 6 a).

Evaluating the total amino acid content, it can be seen that the control showed the highest levels at 5 DAA, reducing and remaining higher than the treatments at 13 and 20 DAA (Figure 3 d), indicating that the amino acids had accumulated in the leaf tissues and this can be considered a sign of plant tolerance to the different environmental stresses. However, most of the treatments remained constant over the course of the DAA, except for the plants treated with RV 1.6 L ha^{-1} + 1.2 L ha$^{(-1)}$FF and RV 1.8 L ha$^{(-1)}$+ 1.5 L ha^{-1} FF, which showed high values at 5 DAA and regressed at 13 and 20 DAA. However, they were lower than the control.

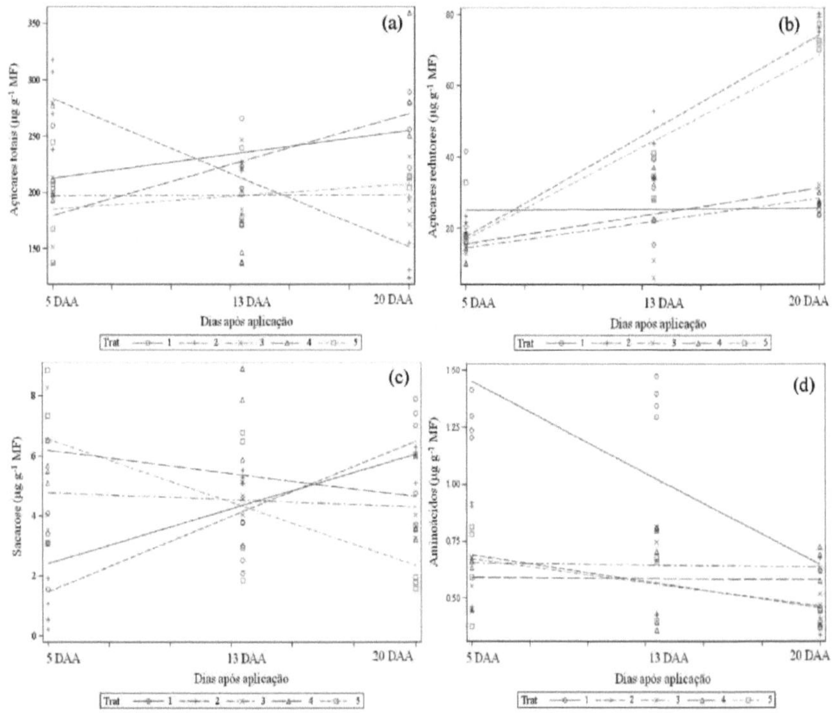

Figure 3- Covariance analyses for total sugars, reducing sugars, sucrose and amino acids µg g^{-1} MF at 5, 13 and 20 days after application of the mixture of plant regulators combined with foliar fertiliser in the berry growth phase (10-12 mm) of the 'Sweet Sunshine' grapevine, in the sub-middle of the São Francisco Valley, Petrolina-PE, 2013.

T1- Witness; T2- RV 1.6 L ha^{-1} + 1.2 L ha^{-1} FF; T3- RV 1.6 L ha$^{(-1)}$+ 1.5 L ha^{-1} FF; T4- RV 1.8 L ha$^{(-1)}$ + 1.2 L ha^{-1} FF; T5- RV 1.8 L ha^{-1} + 1.5 L ha$^{(-1)}$ FF.

For chlorophyll b levels (Figure 4 c), there was a significant interaction between treatments and DAA, where at 5 and 13 DAA, the control treatment had the lowest chlorophyll b levels and the treatment with plant regulators at 1.6 L ha^{-1} + 1.5 L ha^{-1} of foliar fertiliser had the highest values. At 20 DAA, the highest content was observed in the treatment RV 1.6 L ha^{-1} + FF 1.2 L

ha^{-1}, it was also at 20 DAA that all the treatments showed the highest levels. This indicates that the leaves were coming out of the physiological ripeness phase.

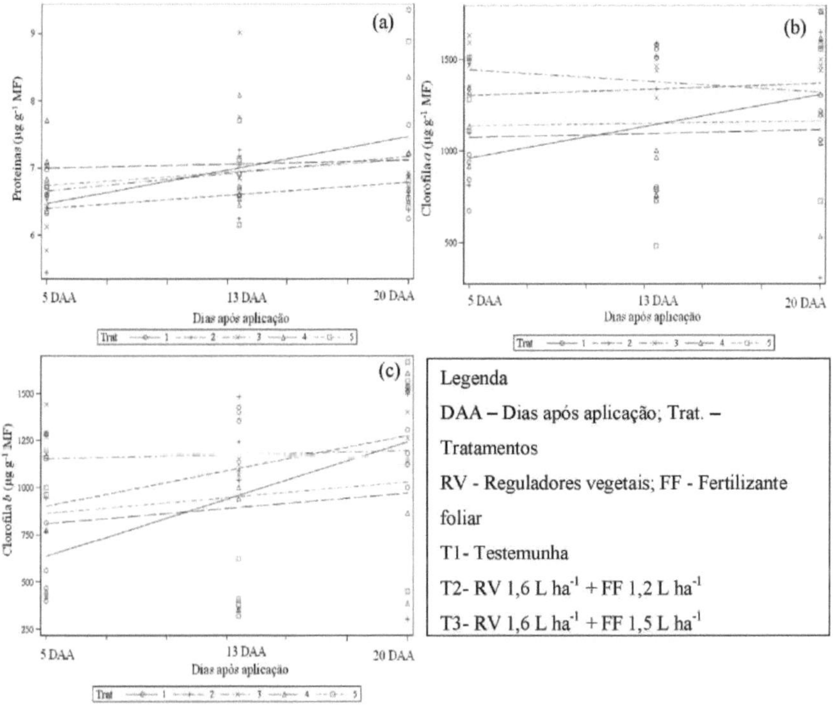

Figura 4. Covariance analyses for leaf protein content, chlorophyll a and b µg g^{-1} MF at 5, 13 and 20 days after application of the mixture of plant regulators combined with foliar fertiliser in the berry growth phase (10-12 mm) of the 'Sweet Sunshine' grapevine, in the sub-middle São Francisco Valley, Petrolina-PE, 2013.Analysing the activity of the antioxidant enzymes, SOD and CAT, at 5 and 13 DAA, it can be seen that the behaviour of the treatments was similar, with the control having the highest activity (Figure 34 a; b). SOD is the enzyme that first acts in defence against oxidative stress and CAT acts in the detoxification of H_2O_2 in plants.

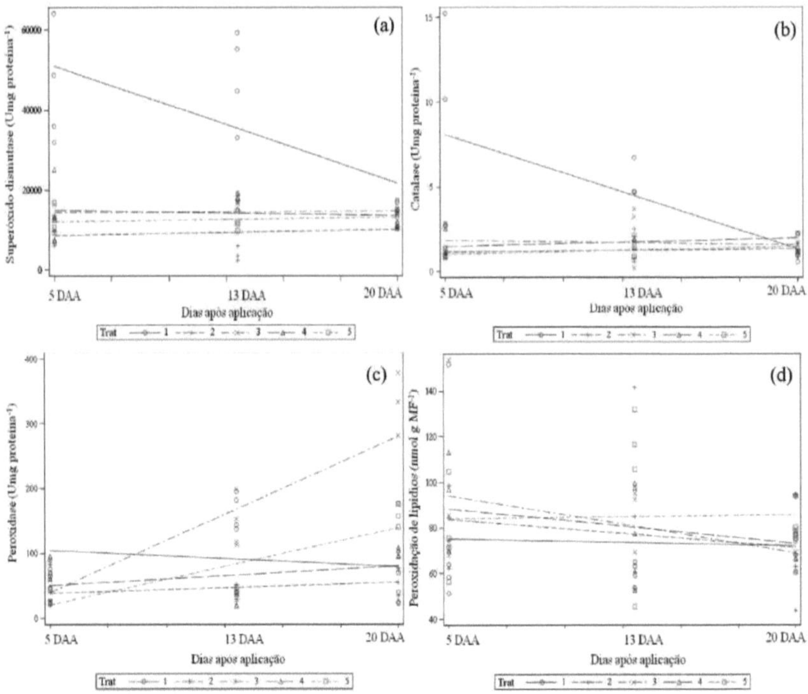

Figura 5. Covariance analyses for superoxide dismutase (U mg^{-1}), catalase (U mg$^{(-1)}$)), peroxidase (Umg^{-1} protein) and lipid peroxidation (nmol g fresh mass$^{-1)}$ activities at 5, 13 and 20 days after the application of plant regulators (VR) combined with foliar fertiliser (FF) in the berry growth phase (10-12 mm) of the 'Sweet Sunshine' grapevine, in the sub-medium São Francisco Valley, Petrolina-PE, 2013.

T1- Witness; T2- RV 1.6 L ha^{-1} + 1.2 L ha^{-1} FF; T3- RV 1.6 L ha$^{(-1)}$+ 1.5 L ha^{-1} FF; T4- RV 1.8 L ha$^{(-1)}$ + 1.2 L ha^{-1} FF; T5- RV 1.8 L ha^{-1} + 1.5 L ha$^{(-1)}$ FF.

There was also a high level of peroxidase activity (Figure 34 c) at 13 and 20 DAA, with the plants treated with the plant regulators at 1.6 L ha^{-1} + 1.5 L ha^{-1} of foliar fertiliser showing the highest activities, acting to combat oxidative stress. Therefore, the treatment acted against oxidative stress, enabling better development of the 'Sweet Sunshine' grapevine plants and, consequently, greater production.

Figure 35 shows that there was no statistical difference for most of the biometric characteristics of the bunches evaluated, except for shoulder length (Fig. 35 d), where all the treatments were superior to the control. However, the difference in production per plant between the control and the above treatment was 2.1 kg, which reflects in the production per hectare an increase of 2993.1 kg, i.e. a gain of approximately 3 tonnes, which is significant for commercial production.

Figures 36c and 37a show that there was only a statistical difference between the treatments for berry diameter

(g) and soluble solids content (°brix), in which the treatment with plant regulators at 1.6 L ha^{-1} + 1.5 L ha^{-1} of foliar fertiliser showed the best results, with a difference of 1.7° brix from the control. Although the above treatment had a higher soluble solids content (15° brix) and lower titratable acidity (0.7 g of tartaric acid 100 g of $pulp^{-1}$), these values are still below the 17° brix required by the market. According to geneticist David W. Cain, who originated this cultivar, increasing the SS content in this variety requires some precautions, such as the appearance of brown spots on the skin of the berries, which intensify as the SS content increases. According to data visualised in the experimental area, belonging to the Labrunier group farm, in the conditions of the São Francisco Valley sub-medium, these spots begin to express themselves with SS content above 18°brix. These spots may be related to an imbalance in the Cu content of white grapevine plants, as Cu acts on the anthocyanin pathway in coloured grapes. High concentrations of this element can cause brown pigmentation in white grape berries.

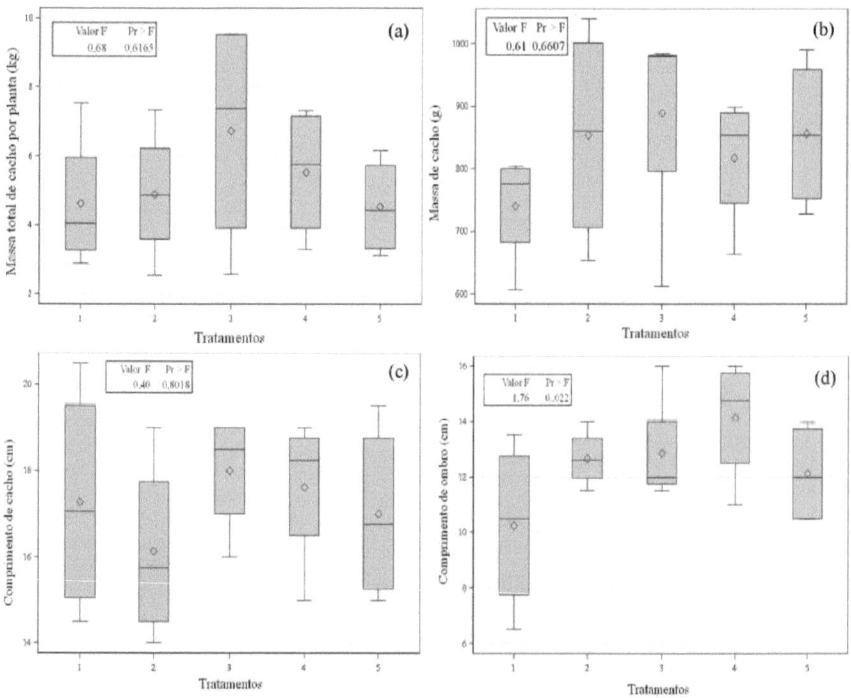

Figura 6. Distribution of total bunch mass per plant (kg), bunch mass (g), bunch length (cm) and shoulder length (CO, cm) after application of a mixture of plant regulators (VR) combined with foliar fertiliser (FF) in the berry growth phase (10-12 mm) of the 'Sweet Sunshine' grapevine, in the sub-middle São Francisco Valley, Petrolina-PE, 2013.

T1- Witness; T2- RV 1.6 L ha^{-1} + 1.2 L ha^{-1} FF; T3- RV 1.6 L $ha^{(-1)}$+ 1.5 L ha^{-1} FF; T4- RV 1.8 L $ha^{(-1)}$ + 1.2 L ha^{-1} FF; T5- RV 1.8 L ha^{-1} + 1.5 L $ha^{(-1)}$ FF.

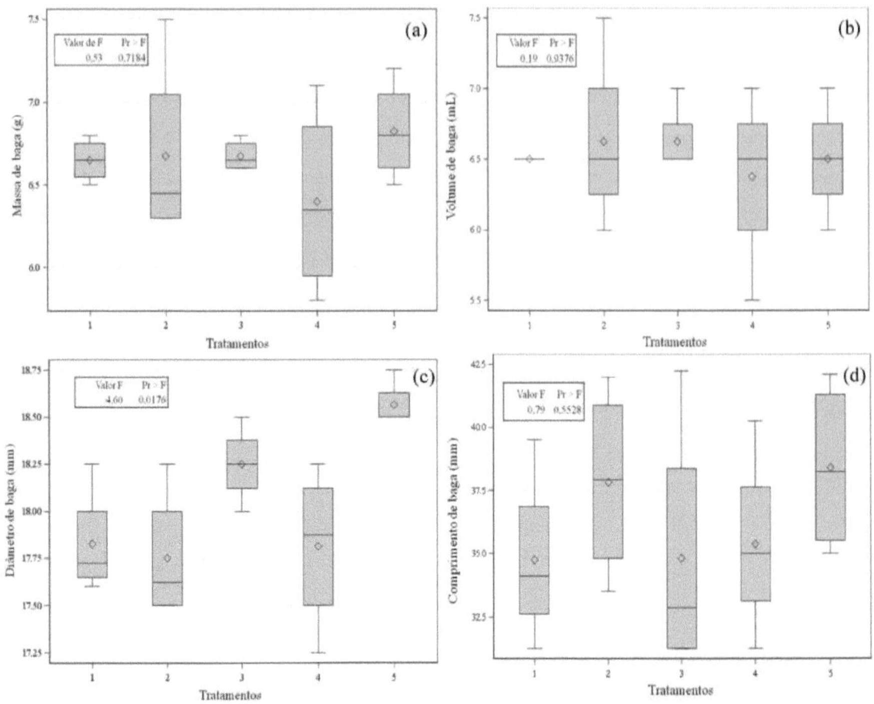

Figura 7. Distribution of berry mass (g), volume (mL), diameter (mm) and length (mm) after the application of a mixture of plant regulators (VR) combined with foliar fertiliser (FF) in the berry growth phase (10-12 mm) of the 'Sweet Sunshine' grapevine, in the sub-middle of the São Francisco Valley, Petrolina-PE, 2013.

(T1- Witness; T2- RV 1.6 L ha^{-1} + 1.2 L ha^{-1} FF; T3- RV 1.6 L $ha^{(-1)}$+ 1.5 L $ha^{(-1)}$FF; T4- RV 1.8 L $ha^{(-1)}$ + 1.2 L ha^{-1} FF; T5- RV 1.8 L ha^{-1} + 1.5 L $ha^{(-1)}$FF).

For the data on the stalk of the bunches (Figures 37 c; 37 d; 38 a, b, c) it can be seen that all the treatments showed roughly similar values and did not differ statistically from one another.

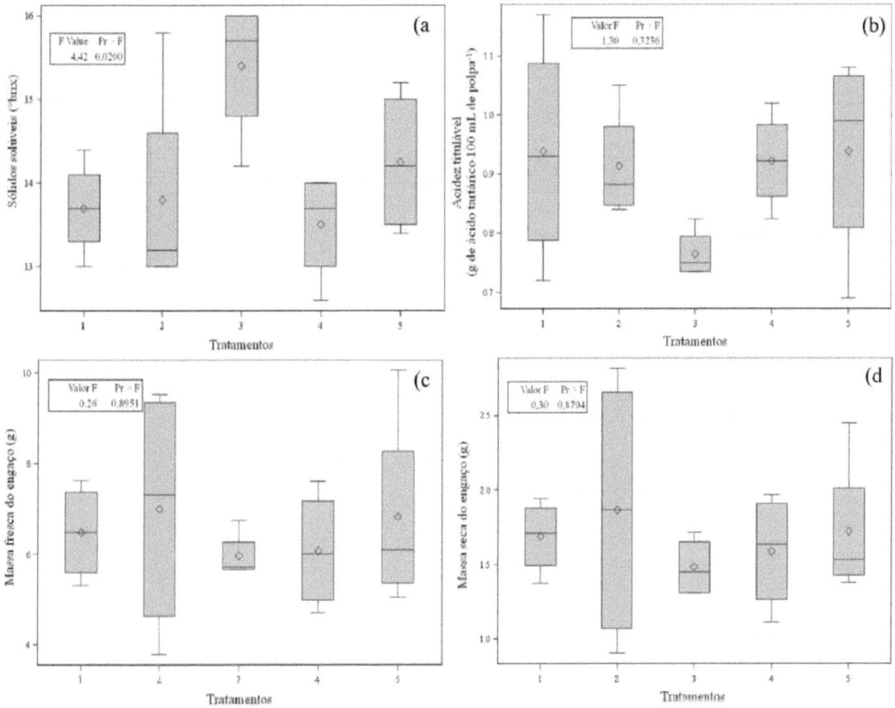

Figura 8. Distribution for soluble solids content (° Brix), titratable acidity (g tartaric acid 100 mL pulp^{-1}), fresh mass (g) and dry mass (g) of the stalk, after application of the mixture of plant regulators (VR) combined with foliar fertiliser (FF) in the berry growth phase (10-12 mm) of the 'Sweet Sunshine' grapevine, in the sub-medium São Francisco Valley, Petrolina-PE, 2013.

(T1- Witness; T2- RV 1.6 L ha^{-1} + 1.2 L ha^{-1} FF; T3- RV 1.6 L ha$^{(-1)}$+ 1.5 L ha$^{(-1)}$FF; T4- RV 1.8 L ha$^{(-1)}$ + 1.2 L ha^{-1} FF; T5- RV 1.8 L ha^{-1} + 1.5 L ha$^{(-1)}$FF).

The condition of the stalk is an important attribute in post-harvest conservation, as it provides a better shelf life, keeping the stalk hydrated for longer, especially in varieties used for export.

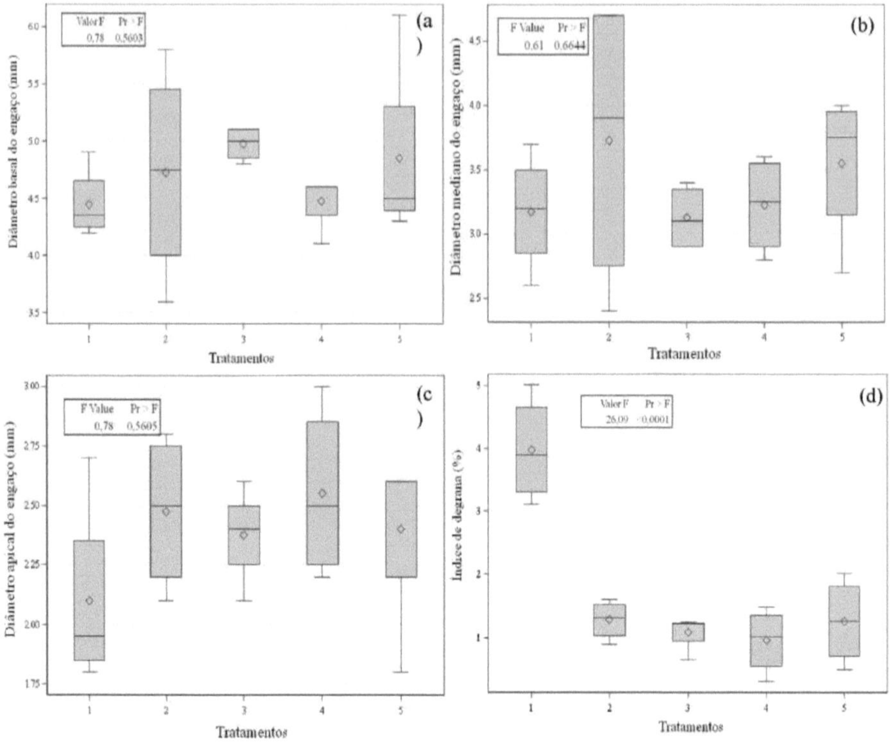

Figura 9. Distribution of the basal (mm), median (mm) and apical (mm) diameter of the stalk, and degranulation index (%) after the application of a mixture of plant regulators (VR) combined with foliar fertiliser (FF) in the berry growth phase (10-12 mm) of the 'Sweet Sunshine' grapevine, in the sub-middle of the São Francisco Valley, Petrolina-PE, 2013.

Analysing the degranulation index (Figure 38 d), it can be seen that all the treatments had a lower degranulation index than the control. This suggests that all the treatments were effective in reducing the abscission of grapevine berries. The reduction in degranulation may be related to auxin, which is the plant hormone responsible for controlling abscission. Therefore, the application of this mixture of plant regulators may have increased the concentration of auxin in the berries, slowing down the process of abscission.Analysing the percentage gain in productivity in relation to the control (Figure 39), it can be seen that treatments T2; T3; T4 and T5 showed a percentage gain in relation to the control of 6.17%; 45.39%; 19.68% and -2.03%, respectively, with the plants treated with the mixture of plant regulators at 1.6 L ha^{-1} + 1.5 L ha^{-1} of foliar fertiliser showing the greatest increase.These effects provided by the treatments on the different variables analysed may be related to the beneficial effects promoted by the nutrients and plant regulators present in the treatments. The mineral nutrients present in the foliar fertiliser used participate in various processes in the plant's physiology, as do the plant regulators.

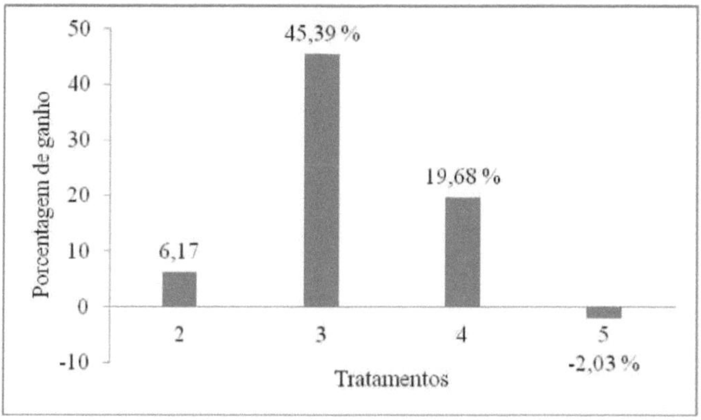

Figura 10. Percentage gain in productivity compared to the control, after application of the mixture of plant regulators (VR) + foliar fertiliser (FF), after application of the mixture of plant regulators (VR) combined with foliar fertiliser (FF) in the berry growth phase (10-12 mm) of the 'Sweet Sunshine' grapevine, in the sub-middle of the São Francisco Valley, Petrolina-PE, 2013.

(T1- Witness; T2- RV 1.6 L ha^{-1} + 1.2 L ha^{-1} FF; T3- RV 1.6 L $ha^{(-1)}$+ 1.5 L $ha^{(-1)}$FF; T4- RV 1.8 L $ha^{(-1)}$ + 1.2 L ha^{-1} FF; T5- RV 1.8 L ha^{-1} + 1.5 L $ha^{(-1)}$FF).

Thus, the combination of plant regulators at 1.6 L ha^{-1} and foliar fertiliser at 1.5 L ha^{-1} promoted changes in the vine's metabolism, increasing production, bunch mass, soluble solids content and reducing acidity and the degranulation index. However, it is suggested that the number of applications of this combination in the first half of the year be increased, with one during the weeping phase and two during the berry growth phase, spaced 10 days apart, as the effects of the treatments were more significant for most of the variables analysed up to 13 days after application.From the results obtained and the conditions under which the experiments were carried out, it can be concluded that:When there is a nutritional and hormonal balance and ideal water content, the Sweet Sunshine grapevine can develop its cycle. Sweet Sunshine grapevine is able to develop its cycle, even under high environmental conditions such as temperature, relative humidity and solar radiation.For the best response to reduce berry degranation in grapevine bunches, we recommend a mixture of plant regulators (IBA + **GA3**+ Kt) at 2.0 L ha^{-1} and 1.6 L ha^{-1} applied alone and/or combined with foliar fertiliser at 1.5 L ha, in the pre-flowering, flowering, fruiting and softening phases in the 1st and 2nd half of the year, respectively.

1. BIBLIOGRAPHICAL REFERENCES

AGRIANUAL. Yearbook of Brazilian Agriculture, São Paulo: FNP, 2012, 482 p.

ALMEIDA, L.P. de.; ALVARENGA, A.A. de.; CASTRO, E.M. de.; ZANELA, S.M.; VIEIRA, C.V. Initial growth of Cryptocaria aschersoniana Mez. plants submitted to levels of solar radiation. **Ciência Rural**, Santa Maria, v.34, n.1, p.83-88, 2004.

ALSCHER, R.G.; ERTURK, N.; HEALTH, L.S. Role of superoxide dismutases (SODs) in controlling oxidative stress in plants. **Journal of Experimental Botany,** Antioxidants and Reactive Oxygen Species in Plants Special Issue, v.53, n.372, p.1331-1341, 2002.

ARAUJO, F.J. Table grape production in Tropical America, In: **International symposium on table grape production**, Anaheim, California, Proceedings, 1994.

ASSIS, J.S. de.; LIMA FILHO, J.M.P. Aspectos fisiológicos da videira irrigada. In: LEÃO, P.C. de S.; SOARES, J.M. **A viticultura no semi-árido brasileiro**. 2. ed. Petrolina: EMBRAPA Semi-Arido. 2000, p.129-142.

ATKIN, O.K.; SCEURWATER, I.; PONTS, T.L. Respiration as a percentage of daily photosynthesis in whole plants in homeostatic at moderate, but not high, growth temperatures. **New Phytologist**, v.174, n.2, p. 367-380, 2007.

AZIZ, A.; LARHER, F.; Osmotic stress induced changes in lipid composition and peroxidation in leaf discs of Brassic nappus L. **Journal of Plant Physiology**, Jena, v. 153, n. 5/6, p.754-762, 1998. BOWLER, C.; VAN MONTAGU, M.; INZÉ, D. Superoxide dismutase and stress tolerance. **Annual Review Plant Physiology and Plant Molecular Biology**, v. 43, p. 83-116, 1992.

BLOKHINA, O.; VIROLAINEN, E.; FAGERSTEDT, K. V. Antioxidants, oxidative damage and oxygen deprivation stress: a review. **Annals of Botany**, London, v. 91, n. 2, p. 179-194, 2003.

BRACKMANN, A.; VIZZOTTO, M.; CERETTA, M. Quality of Dona Zilá and Tardia de Caxias grapes under different storage conditions. **Revista Ciência e Agrotecnologia**, Lavras, v. 26, n.5, p. 1019-1026, 2002.

BRADFORD, M.M. A rapid and sensitive method for the quantification of microgram quantities of protein utilizing the principle of protein-dye binding. **Analytical Biochemistry**, v.72, p.248-254, 1976.

BREUSEGEM, F. V.; VRANOVÁ, E.; DAT, J. F.; INZÉ, D. The role of active oxygen species in plant signal transduction. **Plant Science,** v. 161, p. 405-414, 2001.

CAMARGO, U. A. Impact of Brazilian grape cultivars on the domestic market and potential on the international market. In: BRAZILIAN CONGRESS OF VITICULTURE AND ENOLOGY, 12, 2008, Bento Gonçalves. **Proceedings...** Bento Gonçalves: Embrapa CNPUV, 2008. p. 37-42.

CARMAK, I.; HORST, W.J. Effect of Al lipid peroxidation, superoxide dismutase, catalase and peroxidase activities in root tips of soybean (Glicine max L.). **Physiologia Plantarum**, Copenhagen, v. 834, p. 463-468, 1991.

CAVALCANTI, F. R. et al. Superoxide dismutase, catalase and peroxidase activities do not confer protection against oxidative damage in salt-stressed cowpea leaves. **New Phytologist**, n. 163, n. 03, p. 563-71, 2004.

CHAPPELLE, E.W.; KIM, M.S. Ratio analysis of reflectance spectra (RARS): an algorithm for a remote estimation of the concentrations of chlorophyll A, chorophyll B, and carotenoids in soybean leaves. **Remote Sensing of Environment**, New York, v. 39, p.239-247, 1992.

CHAVES, M. M., PEREIRA, J. S., MAROCO, J., RODRIGUES, M. L., RICARDO, C. P. P., OSÓRIO, M.

L., CARVALHO, I., FARIA, T. & PINHEIRO, C. (2002). How Plants Cope with Water Stress in the Field. Photosynthesis and Growth. Annals of Botany 89: 907-916.

COLAPIETRA, M.; TARRICONE, L.; TAGLIENTE, G. Determinazione delle caratteristiche morfo-produttive dei vitigni di uva da tavola. **L'Informatore Agrario,** Verona. v. 51, n. 49 (supplement), p. 5-34, 1995.

COLL, J. B.; RODRIGO, G. N.; GARCIA, B. S.; TAMÉS, R. S. In: **Fisiología Vegetal**, Madrid: Ediciones Pirámide, 2001, 566p.

COOMBE, B.G. Influence of temperature on composition and quality of grapes. **Acta Horticulturae**, Wageningen, n.206, p.23-35, 1987.

COSTACURTA, A.; ROSELLI, G. Critères climatiques et edaphiques pour l'établissement des vignobles. **Bulletin de Loiv**, Paris, v.53, n.596, p.783-786, 1980.

DAVIES, P. J. **Plant hormones**: biosynthesis, signal transduction, action, 3rd ed. Dordrecht: Kluwer Academic, 2004. 750 p.

DELUC, L. G; GRIMPLET, J.; WHEATLEY, M. D; TILLETT, R. L; QUILICI, D. R; OSBORNE, C.; SCHOOLEY, D.A; SCHLAUCH, K. A; CUSHMAN, J. C; CRAMER, G. R. Transcriptomic and metabolite analyses of Cabernet Sauvignon grape berry development. **BMC Genomics**, v.8, p.429, 2007.

DENG, Y.; WU, Y.; LI, Y.; YANG, M.; SHI, C.; ZHENG, C.A. Mathematical model for predicting grape berry drop during storage. **Postharvest Biology and Technology,** Amsterdam, v. 43, p. 95101, 2007.

DEYTIEUX-BELLEAU, C.; VALLET, A., DONÈCHE, B.; GENY, L. Pectin methylesterase and polygalactoronase in developing grape skin. **Plant Physiology and Biochemistry**, Brussels, v. 46, p. 638-646, 2008.

DÍAZ-RIQUELME, J.; MARTÍNEZ-ZAPATER, J. M.; CARMONA, M. J. Transcriptional Analysis of Tendril and Inflorescence Development in Grapevine (Vitis vinifera L.). **Plos One**, v.9, n.3, 2014. Available at:
http://www.plosone.org/article/info%3Adoi%2F10.1371%2Fjournal.pone.0092339. Accessed 27 Apr. 2014. DOI: 10.1371/journal.pone.0092339.

DIETZ, K.J; HARRIS, G.C. Photosynthesis under nutrient deficiency. In: PESSARAKLI, M. (Org.). Handbook of photosynthesis. New York: Marcel Dekker, 1997. p. 951-975

DINIZ, K.A. **Physiological and biochemical mechanisms of the action of phytohormones on seed germination.** Lavras, 2005. Available at: <http://pt.scribd.com/doc/55110906/1/AUXINAS>. Accessed on: 30 January 2013.

DOKOOZLIAN, N. K. **Table Grape Berry Growth and Development: A Review.** In: Grape Notes. Department of Viticulture and Enology University of California, Davis, 2002, p. 4. Available at: http://cetulare.ucanr.edu/newsletters/March-April_200230429.pdf. Accessed on: 26 Apr. 2014.

ENGEL, V.L.; POGGIANI, F. Study of chlorophyll concentration in leaves and their light absorption spectrum as a function of shading in seedlings of four native forest species. **Revista Brasileira Fisiologia Vegetal,** v.3, n.1,

p. 39-45, 1991.

FANG, W. C.; WANG, J.W.; LIN, C. C.; KAO, C.H. Iron induction of lipid peroxidation and effects on antioxidative enzyme activities in rice leaves. **Plant Growth Regulation,** v.35, p.75-80, 2001.

FARQUHAR, G.D.; SHARKEY, T.D. Stomatal conductance and photosynthesis. **Annual Review of Plant Physiology,** v. 33, p.317-345, 1982.

FERREIRA, E. A.; REGINA, M. A.; CHALFUN, N. N. J.; ANTUNES, L.E.C. Anticipation of harvest for Niágara Rosada grapevine in the southern region of the State of Minas Gerais. **Ciência Agrotecnologia**, Lavras, v. 28, n. 6, p. 1221-1227, 2004

FERRAZ, K. K. F.; SILVA, D. M. 2001. Ecophysiological evaluation of the initial growth of forest species used in the recovery of degraded areas - II. Calliandra calothyrsus Meisn. In: BRAZILIAN CONGRESS OF PLANT PHYSIOLOGY. 8. Abstracts. Ilhéus: Brazilian Society of Plant Physiology. 1 CD-ROM.

FILHO, J. M. P. L.; DANTAS, B. F.; ASSIS, J. S.; SOUZA, C. R.; ALBUQUERQUE, T. C. S. Aspectos Fisiológicos In: SOARES, J. M.; LEAO, P.C.S. (Ed.). **Viticulture in the Brazilian Semi-Arid**: Embrapa, 2009, chap.3, p.73- 108.

FONTES, R. V.; SILVA, D. M. 2001. Ecophysiological assessment of the initial growth of Piptadenia adiantoide (Spreng). Macbr., a forest species used in the recovery of degraded areas. In: BRAZILIAN CONGRESS OF PLANT PHYSIOLOGY. 8, Abstracts. Ilhéus: Brazilian Society of Plant Physiology. 1 CD-ROM.

GASPAR, T.H.; PENEL, C.L.; THORPE, T.; GREPPIN, H. **Peroxidases**: a survey of their biochemical and physiological roles in higher plants. Genève: Université de Genève, 1982. 324p.

GIANNAKOULA, A.; MOUSTAKAS, M.; SYRUS, T.; YUPSANIS, T. Aluminum stress induces up-regulation of an efficient antioxidant system in the Al-tolerant maize line but not in the Alsensitive line. **Environmental and Experimental Botany**, Oxford, v. 67, p. 487-494, 2010.

GIANNOPOLITIS, C. N.; REIS, S. K. Superoxide Dismutases. **Plant physiology,** v. 59, p. 309314, 1977.

GIOVANNINI, E. **Production of grapes for wine, juice and table**. 3. ed. Porto Alegre: Renascença, 2008. 368 p.

HAMADA, K.; HASEGAWA, K.; KITAJIMA, A.; OGATA, T. The relationship between fruit size and cell division and enlargement in cultivated and wild persimmons. **Journal of Horticultural Science & Biotechnology**, Asford, v. 83, n. 2, p. 218-222, 2008.

HAMMOND, J. P.; WHITE, P. J. Sucrose transport in the phloem: integrating root responses to phosphorus starvation. **Journal of Experimental Botany**, Oxford, v. 59, n. 1, p. 93-109, 2008.

HEATH, R. L.; PACKER, L. Photoperoxidation in isolated chloroplasts, I. Kinetics and stoichiometry of fatty acid peroxidation. **Archives of Biochemistry and Biophysics,** v.125, p.189- 198, 1968.

HERNÁNDEZ-MUNOZ, P.; ALMENAR, E.; OCIO, M. J.; GAVARA, R. Effect of calcium dips and chitosan coatings on postharvest life of strawberries (Fragaria x ananassa). **Postharvest Biology and Technology,**

Amsterdam, v.39, n.3, p.247-253, 2006.

HESPANHOL-VIANA, L.; POMMER, C.V.; VIANA, A.P.; CAMPOSTRINI, E. Evaluation of the adherence to the pedicel of the berries of some table grape varieties. **Jornal da Fruta**, Lages, v. 15, n.°192 (November), p. 2-3, 2007.

ADOLFO LUTZ INSTITUTE. Physico-chemical methods for analysing food, ed.IV, 2005. Method 016/IV.

IBRAVIN. **BRAZILIAN WINE INSTITUTE**. Main grape producing regions in Brazil, Rio Grande do Sul, Available at: http://www,ibravin,org,br, Accessed 22 Jun 2012.

IBRAVIN. **BRAZILIAN WINE INSTITUTE**. Promoting grape juice. Available at: <http://www.ibravin.org.br/admin/UPL arquivos/310720091714272.pdf>. Accessed on: 05 Nov. 2014.

IMSANDE, J.; TOURAINE, B. N Demand and regulation of nitrate uptake. **Plant Physiology**, Lancaster, v.105, p.3-7, 1994.

JAIN, A. K. et al. Variation in the sugar accumulation pattern of muscadine grape genotypes. **Proceedings of the Florida State Horticultural Society**, Goldenrod, v. 115, p. 329-336, 2002.

JIFON, J.L.; SYVERTSEN, J.P.; WHALEY, E. 2005. Growth environment and leaf anatomy affect nondestructive estimates of chlorophyll and nitrogen in Citrus sp leaves. Journal of the American Society for Horticultural Science 130: 152-158.

KANELLIS, A. K.; ROUBELAKIS-ANGELAKIS, K. A. Grape, In: SEYMOUR, G,B,; TAYLOR, J.L.; TUCKER, G.A. (eds.). **Biochemistry of fruit ripening**. London: Chapman & Hall, 1993, p.189-234.

KANG, H.; SALTVEIT, M. E. Reduced chilling tolerance in elongating cucumber seedling radicles is related to their reduced antioxidant enzyme and DPPH-radical scavenging activity. **Physiol. Plant**, v.115, p. 244-250, 2002.

KAO, C.H. Differential effect of sorbitol and polyethylene glycol on antioxidant enzymes in rice leaves. **Plant Growth Regulation**, v.39, p.83-89, 2003.

KAR, M.; MISHRA, D. Catalase, peroxidase, and polyphenoloxidase activities during rice leaf senescence. **Physiologia Plantarum**, v.57, p.315-319, 1976.

KAWASHIMA, S.; NAKATANI, M. An Algorithm for Estimating Chlorophyll Content in Leaves Using a Video Camera, **Annals of Botany**, v. 81, p. 49-54, 1998.

KELLER, M, Managing grapevines to optimise fruit development in a challenging environment: A climate change primer for viticulturists. **Australian Journal of Grape and Wine Research**, v.16, p. 56-69, 2010.

KELLER, M.; KUMMER, M.; VASCONCELOS, M.C. Reproductive growth of grapevines in response to nitrogen supply and rootstock. **Australian Journal of Grape and Wine Research**, v.7, p.12-18, 2001.

KENNEDY, J. **Understanding grape berry development**. Winegrowing, Practical Winery and Vineyard, San Rafael. 2002, p. 1-5.

KIM, J.S. et al. Death mechanisms caused by carotenoid biosynthesis inhibitors in green and in undeveloped plant tissues. **Pesticide Biochemistry and Physiology**, v.78, p.127-139, 2004.

KIRKBY, E. A.; ROMHELD, V. **Micronutrients in plant physiology: functions, absorption and mobility**. Technical booklet. International Plant Nutrition Institute. Informações Agronômicas, Piracicaba, n. 118, 2007.

KLUGE, R.A. LCB - 311 **Plant physiology: lecture notes on photosynthesis**. ESALQ/USP. 2004. Available at: http://orion.cpa.unicamp.br/sbfv/ arquivos/aulas/gradOl/06 fotoquímica da fotossintese/fotossínteseKluge. Accessed 27 Apr. 2014

KRAMER, T.; KOZLOWSKI, T. **Physiology of woody plants**. New York, Academic Press, 1979. 811p.

KUMAR, N.; ANAND, K.G.V.; REDDY, M.P. Plant regeneration of non-toxic Jatropha curcas-impacts of plant growth regulators, source and type of explants. **Journal of Plant Biochemistry and Biotechnology**, v.20, p.125-133, 2011.

KUNH, G. B. Description of the plant. In: KUNH, G. B. **Uva para processamento, Produção, Aspectos Técnicos.** Bento Gonçalves: Embrapa Informação Tecnológica, 2003, Chap. 4, p. 24-26. LAMBERS, H.; CHAPIN, F.S.; PONS, T.L. **Plant physiological ecology**. New York: Springer-Verlag, 1998. 540p.

LANTMANN, A. F. **Soya nutrition and productivity with molybdenum and cobalt**. EMBRAPA articles, Coletânea rumos & debates, 2002. Available at: www.embrapa.org.br. Online, Accessed 18 Apr. 2013.

LARKUM, A. W.D.; MICHAEL KÚHL, M. Chlorophyll d: the puzzle resolved. **TRENDS in Plant Science**, v.10, n.8, 2005.

LAWLOR, D.W. Carbon and niotrogen assimilation in relation to yield: mechanismsare the key to understanding production systems. **Journal of Experiment Botany**, v.53, p. 773-787, 2002.

LEÃO, P.C. de S., MAIA, J.D.G. Aspectos culturais em viticultura tropical - Uvas de mesa. **Informe Agropecuário**, Belo Horizonte, v.19, n.194, p.34-40, 1998.

LEÃO, P. C. de S. **Evaluation of the phenological and productive behaviour of six seedless grape varieties in the São Francisco River Valley**. 1999.124f. Dissertation (Master's Degree)-Faculty of Agricultural and Veterinary Sciences, Universidade Estadual Paulista, Jaboticabal, 1999.

LEÃO, P. C. de S.; POSSÍDIO, E. L. de. Management and cultural treatments. In: SOUZA LEÃO, P. C. de (Ed.). **Table grapes: production - technical aspects**. Brasília: Embrapa Informação Tecnológica; Petrolina-PE: Embrapa Semiárido, 2001. ch.10, p. 70-81. (Frutas do Brasil, 13).

LEÃO, P.C.de.S. **Cultivation of the Vine,** 2010. Available: <http://sistemasdeproducao,cnptia,embrapa,br/FontesHTML/Uva/CultivodaVideira_2ed/cultivares, html>, Accessed: 25 Apr. 2013.

LEÃO, P.C. de S; SOARES, J. M.; RODRIGUES, B. L. Main Cultivars. In: SOUZA LEÃO, P.C. de.; SOARES, J.M. **A viticultura no Semiárido**. Ed. Petrolina: Embrapa Semi-Arido, 2009. chap.5, p.151-213.

LEÃO, P. C. de. S.; SILVA, D. J.; SILVA, E. E. G. da. Effect of gibberellic acid, Crop Set Biostimulant and ringing on the production and quality of 'Thompson Seedless' grapes in the São Francisco Valley. **Revista Brasileira Fruticultura,** Jaboticabal - SP, v. 27, n. 3, p. 418421, 2005.

LEBON, G.; WOJNAROWIEZ, G.; HOLZAPFEL, B.; FONTAINE, F.; VAILLANT-GAVEAU, N.;

CLÉMENT, C. Sugars and flowering in the grapevine (Vitis vinifera L.). **Journal of Experimental Botany**, v.59, n.10, p.2565-2578, 2008.

LEE, D.W. Simulating forest shade to study the development ecology of tropical plants: Juvenile growth in three vines in India. **Journal of Tropical Ecology**, Cambridge, v.4, p.281-92, 1988.

LETHAM, D. S.; PALNI L. M. S. The biosynthesis and metabolism of cytokinins. **Annual Review of Plant Physiology**, Palo Alto, v, 34, p, 163-197, 1983

LIMA, S.S. **Growth, Mineral composition and visual symptoms of macronutrient deficiencies in Zingiber Spectabilis Griff plants.** 2009. 84f. Dissertation (Master's) - Federal Rural University of Amazonia, Belém, 2009.

LIMA FILHO, J. M. P.; DANTAS, B. F.; ASSIS, J. S. de; SOUZA, C. R. de; ALBUQUERQUE, T. C. S. de. Physiological aspects, In: SOARES, J, M,; LEAO, P, C, de S, (Ed). **A vitivinicultura no Semiárido brasileiro**, Brasília, DF: Embrapa Informação Tecnológica; Petrolina: Embrapa Semi- Árido, 2009, chap. 3, p. 73-108.

LEUSTEK, T.; MARTIN, M. N.; BICK, J. A.; DAVIES, J. P. Pathways and regulation of sulfur metabolism revealed through molecular and genetic studies. **Annual Review of Plant Physiology and Plant Molecular Biology**, v.51, p.141-165, 2000.

LIN, C.C.; KAO C.H. Effect of NaCl stress on H2O2 metabolism in rice leaves. **Plant Growth Regul**, v.30, p.151-155, 2000.

MAGALHÃES FILHO, J. R. **Photosynthetic responses to changes in root temperature of sweet orange 'Valência' plants grafted on Rangpur lime**. 2009. 50f Dissertation (Master's) - Agronomic Institute of Campinas, 2009.

MAJEROWICZ, N. Photosynthesis. In: KERBAUY, G. B. (Ed.). **Plant Physiology**. Rio de Janeiro: Guanabara-Koogan, 2004. 114-178p.

MALAVOLTA, E. **Manual de nutrição mineral de plantas**. São Paulo: Agronômica Ceres, 2006. 631p.

MALAVOLTA, E. **Elementos de Nutrição Mineral de plantas**: São Paulo: Editora Agronômica Ceres, 1980. 231p

MANDELLI, F.; BERLATO, M. A.; TONIETTO, J.; BERGAMASCHI, H. Grapevine phenology in the Serra Gaúcha. **Pesquisa Agropecuária Gaúcha**, v. 9, n. 1-2, p. 129-144, 2003.

MANDELLI, F. **Meteorological Behaviour and its Influence on the 2009 Harvest in Serra Gaúcha**, 2009. Available at:

http://revistagloborural.globo.com/GloboRural/0,6993,EEC1664546-4529,00.html. Accessed 10 Apr. 2013.

MANGANARIS, G. A.; VASILAKAKIS, M.; DIAMANTIDIS, G.; MIGNANI, I. The effect of postharvest calcium application on tissue calcium concentration, quality attributes, incidence of flesh browning and cell wall physicochemical aspects of peach fruits, **Food Chemistry**, New York, v.100, n.4, p.1385-1392, 2007.

MARENCO, R.A.; LOPES, N.F. **Plant physiology**: photosynthesis, respiration, water relations and mineral nutrition. nutrition. Viçosa: UFV, 439p. 2005.

http://revistagloborural.globo.com/GloboRural/0,6993,EEC1664546-4529,00.html. Accessed on: 10 Apr. 2013.

MARENCO, R.A.; LOPES, N.F. **Plant physiology**: photosynthesis, respiration, water relations and mineral nutrition. Viçosa: UFV, 2011. 439p.

MARSCHNERS, H. **Marschner's Mineral Nutrition of Higher Plants.** 3rd Edition London, New York: Academic Press, 2011, 672 p.

MASCARENHAS, R, J.; SILVA, S. M.; LIMA, M. A. C.; MENDONÇA, R. J. N.; HOLSCHUH, H. J. Characterisation of maturity and quality of Brazilian apirenic grapes in the São Francisco river Valley. **Revista Ciência Tecnologia de Alimentos**, v.32, n.1, 2012.

MELLO, L. M. R de. **Brazilian viticulture**: panorama 2010. Embrapa Uva e Vinho, Available at: http://www.cnpuv.embrapa.br. Accessed 28 Jun. 2012.

MENEZES, A. C. P. **Plant regulators on sprouting, bunch characteristics and productivity of the cv. Itália grapevine in the São Francisco Valley.** 2007. 117f. Thesis (Doctorate) - Universidade Estadual Paulista, Faculdade de Ciências Agronômicas, Botucatu, 2007.

MESSINGER, S. M. et al. Evidence for involvement of photosynthetic processes in the stomatal response to CO2. **Plant Physiologic**, v. 140, n.2, p. 771-778, 2006. doi: http://dx.doi.org/10.1104/pp.105.073676

MIELLE, A.; MANDELLI, F. 2004, Dry pruning of the vine. Available at: < http:// www.cnpuv.embrapa.br/serviços/viticultura/podaseca.htm/ >. Accessed on 20 May 2009.

MILLER, G. L. Use of dinitrosalicylic and reagent for determination of reducing sugar. **Analytical Chemistry**, v.31, n.3, p.426-428, 1959. <http://dx.doi.org/10.1021/ ac60147a030>.

MITTLER, R. Oxidative stress, antioxidants and stress tolerance. **Trends in Plant Science**, v.7, n.9, p.405-410, 2002.

MOLLER, I. M.; JENSEN, P. E.; HANSSON, A. Oxidative modifications to cellular components in plants. **Annual Review of Plant Biology**, v.58, p.459-481, 2007.

MOORE, T. C. Auxins. In: MOORE, T. C. (Ed). **Biochemistry and Physiology of Plant Hormones**. New York: Springer Verlag, 1989. 29-93p.

MOREIRA, E. R. **Effects of gibberellic acid and thidiazuron on bunches and berries of Niagara Rosada grapes**. 2009. 70f. Dissertation (Master's Degree)- Ilha Solteira School of Engineering, Universidade Estadual Paulista, Ilha Solteira, 2009.

MORRISON, J. Bud development in *Vitis vinifera* L. **Bot Gaz**, v.152, p. 304-315, 1991. doi: 10.1086/337894

MOURA, M.F.; TECCHIO, M.A.; TERRA, M.M. ; CIA, P. ; HERNANDES, J.L.; BENATO, E.A.; SIGRIST, M.M.J.; PIRES, E.J.P.; BETTIOL NETO, J.E. Influence of naphthaleneacetic acid and calcium chloride on the reduction of degranulation in Niagara Rosada grapes grown in Jales. In: XIX Congresso Brasileiro de Fruticultura, 2006, Cabo Frio-RJ. Fruits from Brazil: Health for the world. **Lectures and abstracts**. Viçosa-MG : JARD Editora, v. 1, p. 284, 2006.

MOURA, M. S. B. de; TEIXEIRA, A. H. de C.; SOARES, J.M. **Climatic requirements**. In: SOARES, J. M.; LEAO, P. C. de S. A Vitivinicultura no Semiarido Brasileiro. Petrolina: Embrapa Semiarido; Brasilia: Embrapa Informacao Tecnologica. 2009. 37-67p.

MULLINS, M. G.; BOUQUET, A.; WILLIAMS, L. E. **Biology of the Grapevine**. Cambridge University Press, 1992.

NAVES-BARBIERO, C. C. et al. Sap flow and stomatal conductance of two evergreen woody species in campo sujo and cerradão. **Revista Brasileira Fisiologia Vegetal,** v.12, p.119-134, 2000.

OLIVEIRA, M. J. A. et al. Photosynthesis, stomatal conductance and transpiration in pupunheira under water deficiency. **Scientia Agricola,** v. 59, n. 1, p. 59-63, 2002.

PALLIOTTI, A.; CARTECHINI, A. Developmental changes in gas exchange activity in flowers, berries, and tendrils of field-grown Cabernet Sauvignon. **American Journal of Enology and Viticulture**, Davis, v. 52, n. 4, p. 317-323, 2001.

PASCHOLATI, S.F.; LEITE, B. Biochemical mechanisms of disease resistance. **Annual Review of Plant Pathology**, v.2, p.1-51, 1994.

PAULETTO, D.; MOURÃO FILHO, F.A.A.; KLUGE,R.A.; SCARPARE FILHO, J.A. 2001 a. Production and vigour of the vine - Niágara Rosada' related to rootstock. *Pesquisa Agropecuaria Brasileira* 36: 115121

PEREIRA, G. E.; GUERRA, C. C.; MANFROI, L. Vitivinicultura e Enologia. In: SOARES, J. M.; LEÃO, P. C. S. **Vitivinicultura no Semiárido brasileiro.** Petrolina: EMBRAPA Semi-Arid, 2009. 756 p.59

PEREIRA, G.E. **Tempranillo and Petit Verdot gain ground in Semi-Arid wineries**. News, 2009. Available at: <http://www.embrapa.br/imprensa/noticias/2009> Accessed on 5 Nov. 2014.

PETRI, J,L,; PALLADINI, L,A,; POLA, A,C, **Apple tree dormancy and budbreak induction.** In: EPAGRI. **Manual da cultura da macieira**, Florianópolis: EPAGRI, 2002. 261-298 p.

PETRI, J.L.; HAWERROTH, F. J.;LEITE, G. B.; COUTO, M. Chemical thinning in 'Fuji Suprema' and 'Lisgala' apple trees. **Revista Brasileira Fruticultura**, Jaboticabal - SP, v. 35, n. 1, p. 170-182, 2013.

PIMENTEL, C.; PEREZ, A. J. de L. C. Establishment of parameters for the evaluation of drought tolerance in bean genotypes. **Pesquisa Agropecuária Brasileira**, Brasília, DF, v. 35, n. 1, p. 31-39, 2000.

POMMER, C. V.; TERRA, M. M.; PIRES, J. P. Cultivars, improvement and physiology. In: POMMER, C. V. **Uva: production technology, post-harvest, market**. Five Continents. Porto Alegre. 2003. 109-294 p.

POMPELLI, M. F. et al. Photosynthesis, photoprotection and antioxidant activity of purging nut under drought deficit and recovery. **Biomass and Bioenergy**, v. 34, p. 1207-1215, 2010.

PORRA, R. J.; THOMPSON, W. A.; KRIDEMANN, P. E. Determination of accurate extinction coefficients and simultaneous equations for assaying a and b extracted with four different solvents: verification of the concentration of chorophylls standards by atomic absorption spectroscopy. **Biochimic et Biophysica Acta**, Amsterdam, v.975, p.384-394, 1989.

POSSNER, D. R. E.; KLIEWER, W. M. The localisation of acids, sugars, potassium and calcium in developing grape berries. **Vitis**, Siebeldingen, v. 24, p. 229-240, 1985.

PRICE, C. A. et al. Functions of Micronutrients in agriculture. In: MONTVEDT, J. J. et al. (eds). **Micronutrients in agriculture; Zn, Fe, Mo, Cu, B, Mn, Soil Science Society of America**, Madison, 1972. 231-242 p.

PROTAS, J. F. da S.; CAMARGO, U. A. **Vitivinicultura brasileira: panorama sectorial de 2010** / - Brasília, DF: SEBRAE; Bento Gonçalves: IBRAVIN: Embrapa Uva e Vinho, 2011.

QUEIROZ, C. G. S.; ALONSO, A.; MARESGUIA, M.; MAGALHÃES, A. C. Chilling induced changes in membrane fluidity and antioxidant enzyme activities in Coffea arabica L. roots. **Biology Plantarum**, Prague, v. 41, n. 3, p.403-413, 1998.

QUEIROZ, C.G.S.; GARCIA, Q. S.; LEMOS FILHO, J. P. Photosynthetic activity and membrane lipid peroxidation in aroeira-do-sertão plants under water stress and after rehydration. **Brazilian Journal Plant Physiology**, v.14, n.1, p.59-63, 2002.

QUEIROZ-VOLTAN, R. B.; PIRES, E. J. P. A Videira: In: POMMER, C.V. **Uva: do plantio a produção, pós-colheita e mercado.** Porto Alegre: Cinco Continentes, 2003, Chap.1, 185 p.

RAMA DEVI, S.; PRASAD, M. N. V. Copper toxicity in Ceratophyllum demersum L. (Coontail), a free floating macrophyte: response of antioxidant enzymes and antioxidants. **Plant Science**, Limerick, v. 138, n. 2, p. 157-65, 1998.

RAYCHAUDHURI, S. S. The role of superoxide dismutase in combating oxidative stress in higher plants. **The Botanical Review**, New York, v. 66, p. 89-98, 2000.

RICHARDSON, A.D.; DUIGAN, S.P.; BERLYN, G.P. 2002. An evaluation of noninvasive methods to estimate foliar chlorophyll content. New *Phytologist* 153: 185-194

RODRIGUES, A.; ARAUJO, J. P. C.; GIRARDI, E. A.; SCARPARE, F. V.; FILHO, J. A. S. Application of AG3 and CPPU on the quality of 'Itália' grapes in Porto Feliz-Sp, **Revista Brasileira de Fruticultura**, Jaboticabal, v. 33, n. 1, p. 001-007, 2011. http://dx.doi.org/10.1590/S0100- 29452011005000033.

RODRIGUES, B. L. **Effect of gibberellic acid (GA3) on the characteristics of berries and bunches of seedless grapes in the Submedium of the São Francisco Valley-Brazil,** 2008. 96f. Dissertation (Master's) - Bahia State University, 2008.

ROITSCH, T.; BALIBREA, M. E.; HOFMANN, M.; PROELS, R.; SINHA, A. K. Extracellular invertases: metabolic enzyme and metabolic protein. **Journal of Experimental Botany**, v.54, n.382, p.513- 524, 2003.

ROSEN, H. A modified ninhydrin colourimetric analysis for amino acids. **Archives of Biochemistry and Biophysics**, v.67, n.1, p.10, 1957. <http://dx.doi.org/10.1016/0003- 9861(57)90241-2>.

ROITSCH, T.; BALIBREA, M.E.; HOFMANN, M.; PROELS, R.; SINHA, A.K. 2003. Extracellular invertases: metabolic enzyme and metabolic protein. Journal of Experimental Botany 54: 513-524.

ROSS, J.; O'NEILL, D. New interactions between classical plant hormones. **Trends in Plant Science**, v.6, p.2-4, 2001.

RUIZ, V. S. Fitorreguladores, In: **Los Parasitos de la vid: estrategias de proteccion razonada**. 4 ed. Madrid:

Mundi-Prensa, 1998. 303-306p.

SALISBURY, F.B.; ROSS, C. W. **Plant Physiology**, 4. ed. Belmont, CA: Wadsworth publishing, 1992. 682p.

SANTOS, C. E. dos; ROBERTO, S. R.; SATO, A. J.; JUBILEU, B. da. Characterisation of the phenology and thermal demand of 'Cabemet Sauvignon' and 'Tannat' vines for the northern region of Paraná. **Acta Sci. Agron**. Maringá, v. 29, n.3, p. 361-366, 2007.

SANTOS, C. M. G.; MOREIRA, M. M.; RODRIGUES, J. D. Metabolism of 'syrah' grapevine in the northeastern semi-arid region under three water strategies. **Semina: Ciências Agrárias**, Londrina, v. 34, n. 6, supplement 1, p. 3611-3624, 2013.

SCANDALIOS, J. G. Oxidative stress: molecular perception and transduction of signals triggering antioxidant gene defences. **Brazilian Journal of Medical and Biological Research**, Ribeirão Preto, v. 38, n. 7, p. 995-1014, 2005.

SCHADCHINA, T.M.; DMITRIEVA, V.V. Leaf chlorophyll content as a possible diagnostic mean for the evaluation of plant nitrogen uptake from the soil. **Journal of Plant Nutrition**, New York, v.18, p.1427-1437, 1995.

SENTELHAS, P.C. Climatic aspects for tropical viticulture. **Informe Agropecuário**, v. 19, p. 9-14, 1998.

SHARKEY, T.D.; BERNACCHI, C.J.; GRAHAM,D.F.; SINGSAAS, E.L. 2007. Fitting photosynthetic carbon dioxide response curves for C3 leaves. Plant. Cell and Environment 30: 1035-1040

SHIKHAMANY, S.D. 1999. Physiology and cultural practices to produce seedless grapes in tropical environments. In: CONGRESSO BRASILEIRO DE VITICULTURA E ENOLOGIA, 9 Anais... Bento Gonçalves: Embrapa-CNPUV. p.43-48.

SHIM, I. S.; MOMOSE, Y.; YAMAMOTO, A.; KIM, D. W.; USUI, K. Inhibition of catalase activity by oxidative stress and its relationship to salicylic acid accumulation in plants. **Plant Growth Regulation**, v.39, p.285-292, 2003.

SIEGEL, B.Z. Plant peroxidases - an organismic perspective. **Plant Growth Regulation**, v.12, p.303-312, 1993.

SILVA, R. B. C. et al. Water and radiation use efficiency in a mangrove ecosystem in the state of Pará. **Revista Brasileira de Meteorologia**, v. 21, n. 3a, p. 104-116, 2006.

SIMS, D.A. GAMON, J.A. Relationships between leaf pigment content and spectral reflectance across a wide range of species, leaf structures and developmental stages. **Remote Sensing of Environment**, v.81, p. 337-354, 2002.

SMEAL, D.; ZHANG, H. Chlorophyll meter evaluation for nitrogen management in corn. **Communications in Soil Science and Plant Analysis**, New York, v.25, n.9/10, p.1495-1503, 1994.

SILVA, P.C.G da.; CORREIA, R.C. Social and economic characterisation of the grapevine. Embrapa Semiárido. Sistemas de Produção, 1 ISSN 1807-0027 Electronic Version July/2004. Available in: sistemasdeproducao.cnptia.embrapa.br/FontesHTML/Uva/CultivodaVideira/socioeconomia.htm.

Accessed September 2009

SNEDDEN, W. A.; FROMM, H. Calmodulin, calmodulin-related proteins and plant responses to the environment. **Trends in Plant Science**, v.3, p.299-304, 1998.

SOUSA, J. S. I. Uvas para o Brasil. ed.1. Piracicaba: FEALQ, 1996. 791 p.

TERRA, M. M. Nutrition, liming and fertilisation. In: Pommer, C. V. **Uva: production technology, post-harvest, market**. Five Continents. Porto Alegre, 2003. 405-475 p.

SOUZA, J. S. I. de; MARTINS, F. P. **Viticultura Brasileira: principais variedades e suas características.** Piracicaba: FEALQ, 2002. 368p.

SOUZA, A. R. E. **Production and quality of Crimson Seedless grapevine bunches under the action of bioregulators.** 2013. 92f Dissertation (Master's Degree in Agronomy/Irrigated Horticulture) - State University of Bahia - UNEB, 2013.

SOUZA, E. L.; WOYANN, L. G.; AHLERT, R. J.; ALMEIDA, A. M. de.; COSTA de O. A Lipid peroxidation in irrigated rice cultivars subjected to iron stress in hydroponics. **Anais.** CIC 2011. XX Congress of Scientific Initiation. UFPEL. 2011.

SOUZA, E. R. **Physiological behaviour and the influence of defoliation on the production and quality of grapevine bunches cv, Itália in the Vale do Submédio São Francisco**. 2010. 65 f. Dissertation (Master's) - State University of Bahia, Juazeiro, 2010.

SOUZA, E. R.; PEREIRA, M. C.; SANTOS, L. S.; RIBEIRO, V. G.; PIONÓRIO, J. A. A.; ARAÚJO, E. A. Quality of 'Superior Seedless' grapes with applications of benzyladenine combined or not with gibberellic acid. **Revista Caatinga**, Mossoró, v. 23, n. 4, p. 144-148, 2010.

SOUZA, E. R.; RIBEIRO, V. G.; DANTAS, B. F.; LIMA FILHO, J. M. P. Carbohydrate variation in 'Itália' grapevine leaves subjected to different levels of defoliation. **Revista Brasileira Ciências Agrárias**. Recife, v.8, n.4, p.535-539, 2013.

SOUZA, R. P. de.; RIBEIRO, R. V.; MACHADO, E. C.; OLIVEIRA, R. F. de.; SILVEIRA, J. A. G. DA. Photosynthetic responses of young cashew plants to varying environmental conditions. **Pesquisa agropecuaria brasileira**, Brasília , v. 40, n. 8, 2005.

SOUZA, C. R. Leaf and berry metabolism of field-grown grapevines (Vitis vinifera) under different irrigation systems. Doctoral Thesis in Agronomic Engineering, UTLISA, 2004

SPINELLI F, A. W.; RADEMACHER, B. E.; SABATINI, A. C. G.; COSTA, A. Reduction of scabincidence(Venturia inaequalis) in apple with prohexadione-Ca and trinexapac-ethyl, two growth regulating acylcyclohexanediones. **Crop Protection**, v. 29, p. 691-698, 2010.

SRINIVASAN, C.; MULLINS, M. G. Control of flowering in the grapevine (Vitis vinifera L,): formation of inflorescences in vitro by isolated tendrils. **Plant Physiology**, v.61, p.127-130, 1978. SUBASHINI, K.; REDDY, G.M. Effect of salt stress on enzyme activities on callus culture of tolerant and suceptible rice cultivars. **Indian Journal of Experimental Biology,** v.28, p.277-279, 1990.

SUELTER, C.H. Enzimeys activated by monovalent cations. **Science,** v.168, p.789-795, 1970.

TAIZ, L.; ZEIGER, E. **Fisiologia vegetal,** 5ª ed, Trad, de E, R, Santarém, Porto Alegre: Artmed, 2013. 918 p.

TAIZ, L.; ZEIGER, E. 2002. Plant physiology. Sunderland: Sinauer Associates. 690p.

TAIZ, L.; ZEIGER, E. 2009. Plant physiology. Translated by E,R. Santarém. Porto Alegre: Artmed. P.819

TECCHIO, M. A.; TERRA, M. M.; CIA, P.; PAIOLI-PIRES, E. J.; MOURA, M. F.; SANCHES, J.; BENATO, E. A.; HERNANDES, J. L.; VALENTINI, S. R. de. T.; SIGRIST, J. M. M. Effect of naphthaleneacetic acid and calcium chloride in reducing **post-harvest** losses **in 'Niagara Rosada' grapes, Revista Brasileira de Fruticultura,** Jaboticabal, v. 31, n. 1, p. 053-061, 2009.

TEIXEIRA, A. H. de. C.; BASTIAANSSEN, W. G. M.; AHMAD, M. D.; BOS, M. G. Analysis of energy fluxes and vegetation-atmosphere parameters in irrigated and natural ecosystems of semi-arid Brazil. **Journal of Hydrology,** v.362, p.110-127, 2008.

TEIXEIRA, A. H. de. C.; BASTIAANSSEN, W. G. M.; BASSOI, L. H. Crop water parameters of irrigated wine and table grapes to support water productivity analysis in Sao Francisco River basin, Brazil. **Agricultural Water Management,** v.94, p.31-42, 2007.

TEIXEIRA, A. H. de. C.; TONIETTO, J.; PEREIRA, G. E.; ANGELOTTI, F. Delimitation of agroclimatic suitability for grapevines under irrigation in the Brazilian Northeast. **Revista brasileira Eng. Agríc. Ambiental,** Campina Grande, v.16, n.4, 2012.

TEIXEIRA, A. H. de. AZEVEDO, P. V. de. Agroclimatic zoning for European grapevines in the state of Pernambuco. **Revista Brasileira de Agrometeorologia,** Santa Maria, v.4, n.1, p.137- 141, 1996.

TEIXEIRA, A.H.C.; SOUZA, R.A.; RIBEIRO, P.H.B.; REIS, V.C.S.; SANTOS, M.G.L. Aptidão agroclimática da cultura da videira no Estado da Bahia, Brasil. **Revista de engenharia Agrícola e ambiental,** Campina Grande, v.6, n.1, 2002.

TERRA, M.M.; PIRES, E.J.P.; NOGUEIRA, N.A.M.; POMMER, C.V. **Tecnologia para produção de uva Itália na região noroeste do Estado de São Paulo.** Campinas: CATI, 1998, (Technical Bulletin, 97).

TERRA, M. M. Nutrition, liming and fertilisation. In: Pommer, C. V. **Uva: production technology, post-harvest, market.** Five Continents. Porto Alegre, 2003. 405-475 p.

TESSMANN, D J; DIANESE, J C; GENTA, W; VIDA, J B; MAYDEMIO, L L. 2004. Grape rust (Phakopsora euvitis), a new disease for Brazil. Phytopathology of Brazil 29: 338

THIMM, O.; BLASING, O.; GIBON, Y.; NAGEL, A.; MEYER, S.; KRUGER, P.; SELBIG, J.; MULLER, L.A.; RHEE, S.Y.; STITT, M. MAPMAN: a user-driven tool to display genomics data sets onto diagrams of metabolic pathways and other biological processes. **Plant Journal,** v.37, p.914-939, 2004.

THORNBER, J.P. Chlorophyll-proteins: light-harvesting and reaction centre components of plants. **Annual Review of Plant Physiology,** California, v.26, p.127-58, 1975.

TONDATO, C. **Characterisation of table grape marketing channels in the north-western region of the state of São Paulo**, 2006. 151f Dissertation (Master's) - Federal University of Mato Grosso do Sul, Campo Grande, 2006.

VASCONCELOS, M. C.; GREVEN, M.; WINEFIELD, C. S.; TROUGHT, M.C.T.; RAW, V. The Flowering Process of Vitis vinifera: A Review. **Am. J. Enol. Vitic.**, v.60, n.4, p. 441-434, 2009.

VIEIRA, C.R.Y.I.; PIRES, E.J.P.; TERRA, M.M.; TECCHIO, M.A.; BOTELHO, R.V. Effects of gibberellic acid and thidiazuron on fruit and must characteristics of 'Niagara Rosada' grapes. **Revista Brasileira Fruticultura**, Jaboticabal, v. 30, n.1, 2008.

VIEIRA, E. L.; SOUZA, G. S, de.; SANTOS, A. R. dos.; SANTOS SILVA, J. dos. **Manual of plant physiology**, São Luís, MA: EDUFMA, 2010, 230p, Available at:<http://books,google,com,br/books?hl=ptBR&lr=&id=enzO_ItTcvMC&oi=fnd&pg=PA1&dq=%27%27Manual+de+fisiologia+vegetal+2010%22&ots=S4o7J56e1b&sig=E3hIlqlch3EbzebSXSYOjtoPtJk#v=onepage&q="Manual%20de%20fisiologia%20vegetal%202010%22&f=false >. Accessed 20 Nov. 2013.

VIEIRA, A.J.D. 1999. Branch growth of Vitis vinifera L. cv. Thompsom Seedless. Brazilian Journal of Agrometeorology 7: 45-52.

VON CAEMMERER S. 2000. Biochemical Models of Leaf Photosynthesis. CSIRO Publishing, Collingwood, Victoria, Australia. p. 1- 165.

VON CAEMMERER, S.; FARQUHAR, G. D. Some relationships between the biochemistry of photosynthesis and the gas exchange of leaves. **Planta**, Berlin, v. 153, n. 4, p. 376-387, 1981.

VON ELBE, J. H. Colourants. In: FENNEMA, O.W. **Química de los alimentos**. 2.ed. Zaragoza: Wisconsin - Madison, 2000. Chap.10, 782-799 p.

WHATLEY, J.M.; WHATLEY, F.R. **A luz e a vida das plantas**. São Paulo, EPU EDUSP, 1982. 101p. (Temas de biologia, 30).

WHITHAM, F. H.; BLAYDES, D. F.; DEVLIN, R. M. **Experiments in Plant Physiology**. New York, D. Van Nostrand Company, 1971. 55-8 p.

WINKLER, A.J.; COOK, J.A.; KLIWER, W.M. **General viticulture**. 2.ed. Berkeley: University of California Press, 1974. 710p.

YUAN, R.; GREENE, D.W. Benzyladeiúne as Chemical thinner for 'McIntosh' apples. I. Fruit thinning effect and associated relationship with photosynthesis, assimilate translocation, and non- structural carbohydrates. **Journal of American Society for Horticultural Science**, Alexandria, n.125, p.169-176, 2000.

ZHÃO, F. J.; HAWKESFORD, M. J.; MCGRATH, S. P. Sulphur assimilation and effects on yield and quality of wheat. **Journal of Cereal Science,** London, v. 30, p. 1-17, 1999.

ZHANG, Y. L.; ZHANG, R.G. Effects of ABA content on the development of abscission zone and berry falling after harvesting of grapes. **Agricultural Sciences in China**, China, v. 8, n.1, p.59-67, 2009.

MIX
Papier aus verantwortungsvollen Quellen
Paper from responsible sources
FSC® C105338

Printed by Books on Demand GmbH, Norderstedt / Germany